Acknowledgements

The author would like to thank the following individuals for their help in making this book possible:

Jim Cofield

Donald Hudson

J. D. Cook

Sherman Clark

BROWN'S ALCOHOL

MOTOR FUEL COOKBOOK

by Michael H. Brown

DESERT PUBLICATIONS

BROWN'S
ALCOHOL MOTOR FUEL
COOKBOOK

by

Michael H. Brown

ISBN 0 - 87947 - 300 - 2

DESERT PUBLICATIONS
Cornville, Arizona 86325

Publishers Note:

Being one who has been concerned with our energy problem for some time, I'm very enthusiastic about the potential of this book. Contained within these covers is the foundation for a totally energy independent America. Although several years of painstaking research and development went into this book by the author, Michael H. Brown, I feel it is only the tip of the iceberg for alcohol, not only for motor fuel but for other fuel requirements such as home heating.

Alcohol is a replenishable energy source, made from nearly all organically grown crops. The grain used for making alcohol is a better livestock feed after alcohol is distilled from it than it was before. It, also, burns clean and helps insure an ecologically safe enviornment.

After initially reading Brown's manuscript my curiosity was aroused as to the capabilities of alcohol as a motor fuel. Although enthused, I was still apprehensive to a large degree.

Upon arriving home that evening, I went directly to my old faithful, gasoline powered, lawnmower and proceeded to empty the fuel tank and carburetor of gasoline. The only alcohol available was a bottle of 70% isopropyl (rubbing) alcohol from the medicine cabinet. After emptying the contents into the fuel tank I adjusted the carburetor according to Brown's directions. Five pulls on the rope and it was running. After a bit more fine tuning of the carburetor, I mowed my lawn. IT WORKS! Son-Of-A-Gun.

Larry Moore
Publisher

WARNING!

FOR CENTURIES ALCOHOL HAS BEEN TAXED, AND REGULATED BY MOST GOVERNMENTS OF THE WORLD INCLUDING OURS. THE READER OF THIS BOOK IS ADVISED TO BECOME COGNIZANT OF ALL FEDERAL, STATE AND LOCAL LAWS CONCERNING THE MAKING OF ALCOHOL FOR MOTOR FUEL.

THE READER IS FURTHER CAUTIONED ABOUT THE DANGERS INHERENT WITH THE HEATING OF ANY LIQUID IN A CLOSED CONTAINER. FAULTY EQUIPMENT OR CARELESS OPERATION CAN BE EX—TREMELY DANGEROUS. TO GRAPHICALLY DRIVE THE POINT HOME WE SHOW BELOW WHAT'S LEFT OF A STEAM TRACTION ENGINE THAT "BLEW," TAKING THREE LIVES.

Table of Contents

Instructions for

Using This Manual

This manual is not, nor was it ever intended to be, a literary masterpiece. All it was designed to do was impart information. However, information is many times a slippery and elusive thing. What is crystal clear to one man may be unadulterated confusion to the next.

Much of the confusion often stems from haste or carelessness. There are a couple of ways to avoid such confusion with this book.

The first is to study the material carefully before you run out and attempt anything. If this was a college textbook I could almost guarantee a "once over lightly" would cause you to flunk the course.

The second is to break your projects down into increments.

If you haven't learned how to take a carburetor apart, remove the jets, drill jets out oversize, stick the carburetor back on the vehicle and get it to work again (on alcohol) you have no business pulling the heads off and milling them, advancing the timing, and so on. Perfect each phase of the operation by itself. And if your engine isn't running well on gasoline correct that problem before you even begin to work on your carburetor.

The same procedure holds true for the fermenting and distilling processes.

If your still will not distill water you have no business putting fermented material in it. It won't distill either.

If you put beer in your still, distill it, and the percentage of alcohol in the fluid that has been distilled doesn't increase then you had better stop right there and figure out what you did wrong when you built the still or what mistake you made in your cooking temperature.

If you can't get rotten fruit to ferment and then distill it, you had better lay off starch crops until you get your act together.

If you can't get a proper reaction with your malt or your yeast, then back up, reread the appropriate sections, and try again. There is always one sure solution when you hit an obstacle and can't seem to make any further progress:

Go back and follow the directions.

Historical Background

Taxes on and problems with alcohol are nothing new. As long ago as 1626 the English king tried to put a tax on spirits. The measure caused such a violent rebellion that the English Parliament declared it unconstitutional. King Charles tried to pass the excise tax on alcohol again in 1641, which may have been how and why he irritated an English farmer by the name of Oliver Cromwell and subsequently lost his kingdom.

The English monarch tried to tax alcohol again in 1734 and was again unsuccessful. Troops had to be called out to restore order in the streets.

It wasn't until March 3, 1791 that an excise tax on alcohol was passed. Not in England, but in the United States. The American Farmer was enraged. A tax on tea was nowhere as devastating as a tax on alcohol. A backwoods farmer could load four bushels or thereabouts of corn or wheat on a mule and take it to market. Obviously there was no profit in feeding one mule for several hundred miles to get four (by-then moldy) bushels of corn to market. By distilling the grain first the same mule could be loaded with the equivalent of three dozen bushels of grain, the skill of the distiller would add to the value of the product, and there would be no problem with mold or mildew.

What the tax did was deprive the farmer of his profit. In

3

addition, at that time alcohol was the farmer's currency, a monetary unit, much in the same fashion that the OPEC nations use oil today as a "medium of exchange." The federal government wanted the tax paid in cash, of which the farmers had none. Unless, of course, they sold their whiskey first. Theirs was basically a barter economy.

Revenue agents were tarred and feathered on a regular basis and by 1794 the farmers mounted what is now known as the Whiskey Rebellion. The Rebellion collapsed when the farmers ran head on into federal troops who outnumbered them 15,000 to 5,000, a larger force than they had ever had to contend with under George III.

The tax on alcohol was repealed under Thomas Jefferson and reinstated under Lincoln to finance the Union half of the War Between the States.

The tax for alcohol unsuitable for drinking (denatured) was repealed by Congress on June 7, 1906. Alcohol has been economically feasible as a motor fuel ever since then.

To quote one authority:

"Recently in this country a widespread interest has developed in the possibilities of alcohol as a fuel. The matter is important to the agricultural interests of the country, both because alcohol is manufactured from various products of the soil and because of small liquid-fuel engines is very common among agriculturists."

> **U.S. Department of Agriculture**
> **Bulletin 191**
> **Issued September 11, 1907**

This book will show you how.

Engine Basics

Any engine must have three things present to operate satisfactorily.

The first is fuel. Obviously that is what this entire book is concerned with, one type of fuel, its production, and utilization. However, a number of other substances have been used as fuel in the past and still others will no doubt be used in the future.

Modern engine fuels have varied more than most people imagine. A diesel engine was originally designed to utilize coal dust. Rudolf Diesel attempted to solve the problem of great piles of powdered coal sitting outside the mines in Germany that no-one had any use for by the invention and development of his engine. The diesel engine was soon found to work better on petroleum fuel and the Chinese ran a large number of diesel engines on vegetable oil before World War II.

The engine found in your car is even more versatile, probably because its predecessors were built upon the erroneous assumption that power can be extracted from gasoline by compressing it, igniting it, and causing it to expand. But gasoline doesn't expand in the common every day usage of the word. It explodes. The power utilized from the typical gallon of gasoline can best be compared to a train of gunpowder spread out on the ground and lit with a match. Most of the energy is expended uselessly into the air. Take the same gunpowder and put it in a cartridge, put the bullet in the chamber of a rifle and pull the trigger: the amount of energy utilized is literally enormous.

The concept of expanding gas originated with the steam engine. That is, water heated to boiling would expand to 1700 times its original volume and the pressure thus created was utilized to drive a piston or do other mechanical work. One enterprising fellow by the name of Jacob Perkins used the principle of expanding steam to build a machinegun in 1800 that would crank out 1,000 rounds a minute. As usual, nobody was interested.

The first engine ever marketed commercially was in fact a converted steam engine with slide valves run on illuminating gas. The man who sold it (or them, whichever you prefer) was a man named Lenoir who began production in 1860. The poppet valve and camshaft was available pretty much as we know them had he wanted to use them since James Watt had been using them on his steam engines since 1800.

The evidence that our modern gas engine is a steam engine with ignition and carburetion added is rather plain. To convert a modern so-called gasoline engine back to steam all that is required is the removal of the carburetor and the installation of a steam line from a boiler in its place, replace the camshaft gear with one the same diameter as the crankshaft timing gear, and install a shorter timing chain. What you do in such a conversion is simply eliminate the ignition and compression strokes. The fact that the older gasoline engines, up until the 1940's, were long-stroke engines reinforces this contention. Gasoline engines since then have almost always been short stroke, longer lasting, and more economical.

After Dr. Otto built the first four-stroke engine in 1876 on the principle of expanding gas and internal combustion (steam is external combustion which means that the energy required to drive the engine is created outside the engine itself) a number of different individuals tried a number of different substances to drive the piston down (or up, on some of the earlier models) on the power stroke.

Gunpowder was one. Most folks weren't very successful.

Grain dust was another. On the pre-1900 large stationary

one cylinder engines it sometimes worked quite well. Grain dust is highly explosive. This explosive quality of grain dust was quite well illustrated several years ago by the explosion of several grain elevators in this country. OSHA inspectors had ordered the windows of the grain elevators closed to keep the grain dust from "polluting" and in so doing created a number of bombs waiting for a spark to set them off.

By 1900 internal combustion engine fuel had almost all evolved to the use of either gasoline in this country or alcohol overseas. The German government awarded prizes for alcohol-fueled efficiency improvements starting in 1894 with engines that were for the most part 6% efficient. By 1900 efficiency of many alcohol engines was approaching 30%. At the time many German farms had their own distilleries capable of producing anywhere from 50 to 500 gallons of alcohol a day so there was plenty of fuel to experiment with.

Up until World War II practically all vehicles, including locomotives, in New Zealand and the Phillipines were alcohol powered. Gasoline was too plentiful and cheap in the United States for alcohol to be really competitive with it until just recently.

So much for fuel. How you get it into your engine will be covered in the chapter on carburetors.

The next item, on an internal combustion engine, to make things happen is compression. The piston, acting as a suction pump, draws fuel into the cylinder, the intake valve closes, and the piston moves up to compress the fuel and air drawn into the cylinder. A partial application of the gunpowder in the cartridge idea.

But only partial.

As the piston compresses the volume of air and atomized liquid fuel in the cylinder the pressure created causes the temperature of the mixture to rise. At a pressure required to attain 1800 degrees F. at spark ignition a carbon oxygen explosion takes place, the "ping" you hear when low grade hydrocarbon fuels are used that is caused by the cylinder walls ringing from shock. Modern gasoline is "doped" with tetraethyl lead and other substances to prevent the ignition of carbon. Detonating fuel expands at the rate of 5,000 feet per second, 2,000 feet per second faster than nitroglycerine dynamite. Your car engine is designed to use only the hydrogen part of hydrocarbon fuel. Hydrogen expands at a rate of between 25 and 75 feet per second. Prior to the discovery and application of lead in 1925, four to one was about as high a compression as anyone could go. By "four to one compression" is meant if the top of the piston is five inches below the cylinder head to bottom dead center of the crankshaft revolution and capable of four inches upward movement to compress the air-fuel mixture. If the same piston was capable of four and half inches of upward movement the compression ratio would be nine to one. That is, 4.5 to 5 or double it to 9 as compared to 10 (subtract the length of the piston stroke from the length — or volume — of the cylinder).

By 1935 compression ratios were in the neighborhood of 7 to 1 and by 1968 many vehicles had engines with 10.5 to 1 ratios. General Motors put out a booklet in 1935 telling us that if compression ratios could be raised from the then 7 to 1 to 10.5 to 1, the pollution equivalent to that blanketing 17 states would not exist. About 1969 all the lawmakers and bureaucrats got into the act and ordered Detroit to lower compression ratios in order to "pollute less."

Ever see an engine designed by a lawyer?

Higher octane fuels (more dope) had to be used in such engines. Higher octane does not mean the gasoline is capable of more power on a raw energy basis. Just the opposite is true. Alcohol has only two-thirds the energy potential of gasoline. Yet alcohol has an octane rating of 115. "Premium" gasoline is normally rated around 100 octane and white gas, suitable only for engines with a 6 to 1 compression ratio and less, has an octane rating of 60.

Ignition is the third element necessary to make an engine work. Something has to ignite the air-fuel mixture after compression, much in the way a primer cap, percussion cap, or flint on a flint-lock rifle creates a spark to ignite gunpowder in a rifle.

The spark must occur at the proper time, hence the installation of distributors to distribute the spark at the proper time (called timing), vacuum lines to the intake manifold to advance the timing when greater vacuum is created by more air and fuel rushing into the engine, a battery to store the electricity needed to create a spark in the proper time and place, a generator to recharge the battery, a regulator to keep the battery from overcharging and blowing up, and various other parts, pieces, and wires.

The steam engine needed only a source of heat for the water in its boiler.

The diesel ignites by compression alone though it normally has a battery merely to turn the engine over until the diesel fuel ignites.

A steam boiler can be heated by alcohol used in a bunsenburner type device or by the "fuel bricks" (alcohol and wax) supplied with model and toy steam engines.

The use of alcohol in diesel is questionable. Most researchers report somewhat negative results running diesel on alcohol.

In the opinion of this author, the use of gasoline in passenger car and small truck engines is questionable.

We should have started with alcohol in the first place.

Note: For you picky scientific types, octane means resistance to detonation. That is, the higher the octane is the harder it is to detonate. The correlation between energy and octane is by no means an absolute one.

SO SIMPLE EVEN A CHILD CAN DO IT. THE AUTHOR'S SON IS SHOWN RUNNING HIS 3hp TECUMSEH-POWERED MINI-BIKE ON RUBBING ALCOHOL.

THE AUTHOR IS SHOWN RIDING HIS 1949 HARLEY-DAVIDSON WHICH IS POWERED ON ALCOHOL.

The Carburetor

The first modification required to run an engine on alcohol consists of changing a tiny piece of brass called a jet. All carburetors designed to use liquid fuel have a jet, in the case of carburetors designed to work on V-8 engines there will normally be jets in multiples of even numbers. For example, a "two-barrel" carburetor will normally have two jets and two venturi tubes, a "four-barrel" carburetor will normally have four jets and four venturi tubes, and so on.

The venturi tube is a restricted passage in the body of the carburetor which causes the air to speed up, lose some of its pressure, and creates a partial vacuum. Since gasoline must be vaporized before it can be ignited and liquid cannot exist in a vacuum the partial vacuum created in the venturi tube helps vaporize the gasoline. It is possible to build a carburetor without a venturi tube. In the early 1900's some manufacturers did.

If you are not used to working on cars, let alone carburetors, there are a couple of procedures that may allow you to get the job done without having a bucket of parts left over.

The first is to go to an auto supply parts house and buy a rebuilding kit for your carburetor. There will be an exploded-view picture (called a sub-assembly drawing by machine tool operators) in the package that you can use as a guide.

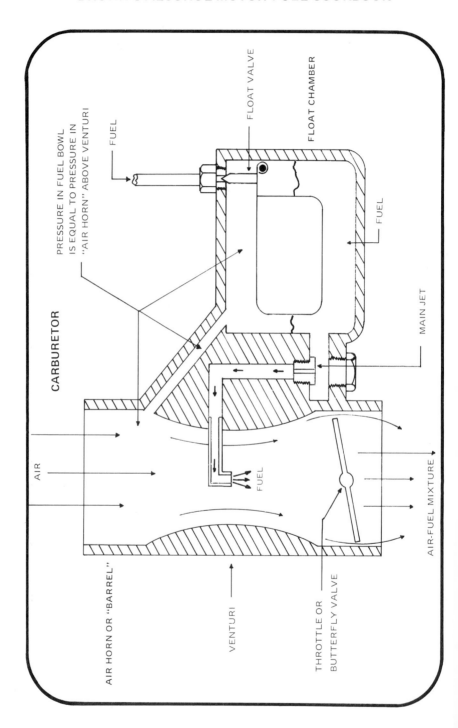

CARBURETOR

AIR

AIR HORN OR "BARREL"

VENTURI

THROTTLE OR BUTTERFLY VALVE

FUEL

AIR-FUEL MIXTURE

MAIN JET

FLOAT CHAMBER

FLOAT VALVE

FUEL

FUEL

PRESSURE IN FUEL BOWL IS EQUAL TO PRESSURE IN "AIR HORN" ABOVE VENTURI

The second is one that works for the bungling amateur every time. Well, almost every time.

Simply take good clear polaroid snapshots of the carburetor linkage as you disconnect it from the car. Follow the same procedure when you disassemble the carburetor on the kitchen table. Number the photos on the back as they come out of the camera. You can do the same thing with a pencil and paper if a polaroid is not available. Too bad if you flunked art class.

To make reassembly easier number the pictures in reverse order as you take them. Start with an arbitrary number such as 100 and number them as they come out of the camera 100, 99, 98, 97, etc. Saves a lot of confusion when it comes time to put the photos in order for reassembly.

If all you want to do is be able to get from point A to point B on alcohol all that is required is a 40% enlargement over stock size of the main jet. That is, just make a 40% larger hole in it. The hole in the main jet that the gasoline flows through is normally in the .050 to .060 range. Just multiply the size of the main jet by 40% and add the product of the multiplication to the size of the main jet. Then find a drill bit to correspond to that dimension.

At this point the purist will be setting up his drill press but a handheld drill will work almost as well.

You can usually get the exact size jet your carburetor requires from the specification charts in most automotive repair manuals. Otherwise you had better take your jet (or jets) to a machinist and have him measure it (or them) for you.

There are several different schools of thought as to why the jet(s) must be bored out to accomodate an alcohol-powered engine.

The first is that alcohol requires a fuel/air or air/fuel ratio of 9 to 1 to ignite in the engine. That is, nine pounds of air for every one pound of alcohol. Gasoline requires a 15 to 1 ratio of air to fuel. The difference between alcohol and gasoline is roughly 40%. The alcohol carries oxygen in the fluid where gasoline does not, which may explain why alcohol requires less air to detonate or ignite.

The second and third appear to be interrelated. Water

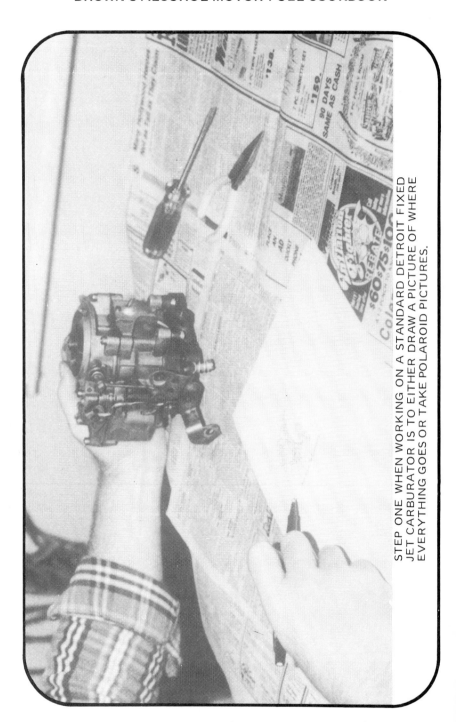

STEP ONE WHEN WORKING ON A STANDARD DETROIT FIXED JET CARBURATOR IS TO EITHER DRAW A PICTURE OF WHERE EVERYTHING GOES OR TAKE POLAROID PICTURES.

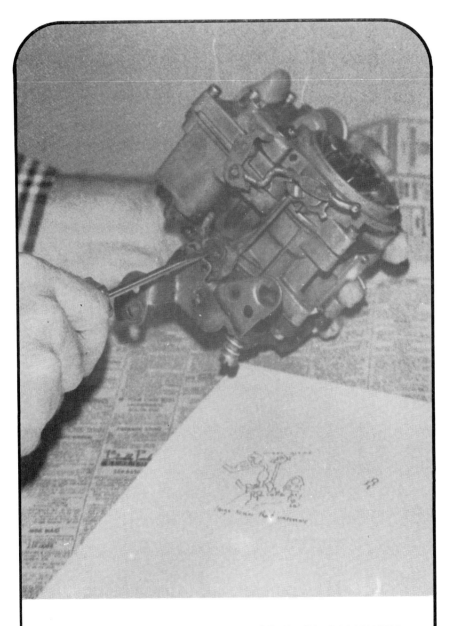

ALWAYS DISCONNECT THE LINKAGE FIRST. CARBURETOR USED IN THIS SEQUENCE IS A 2 - BARREL ROCHESTER FROM A 307 CHEVY V8.

UNSCREW AND SEPARATE THE TWO HALVES OF THE CARB-
URETOR. ON A HOLLEY THE FLOAT COVER UNSCREWS,
WHICH IS A LOT SIMPLER.

TAKE THE TOP HALF (left hand) AND SET IT ASIDE.

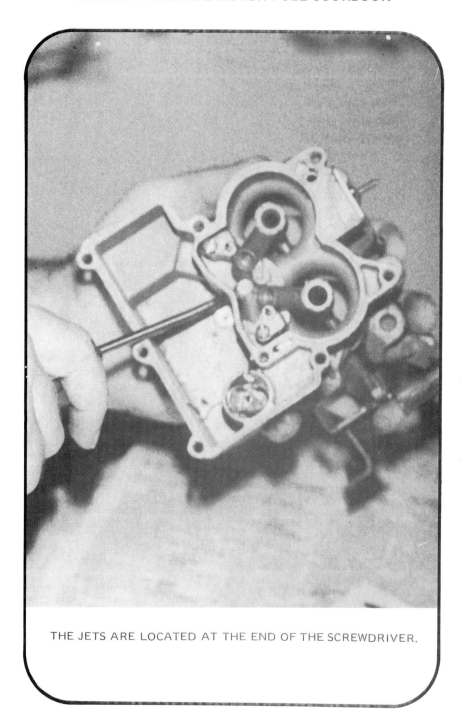

THE JETS ARE LOCATED AT THE END OF THE SCREWDRIVER.

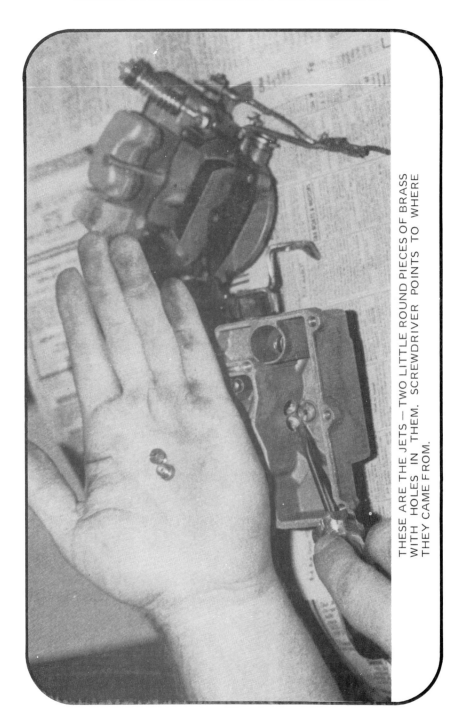

THESE ARE THE JETS — TWO LITTLE ROUND PIECES OF BRASS WITH HOLES IN THEM. SCREWDRIVER POINTS TO WHERE THEY CAME FROM.

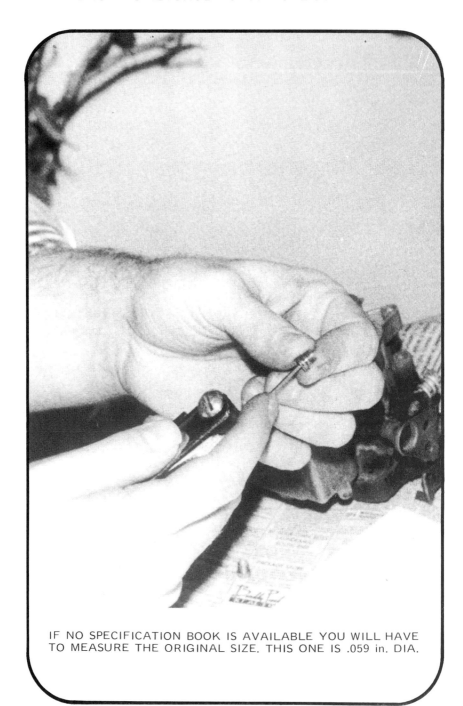

IF NO SPECIFICATION BOOK IS AVAILABLE YOU WILL HAVE TO MEASURE THE ORIGINAL SIZE. THIS ONE IS .059 in. DIA.

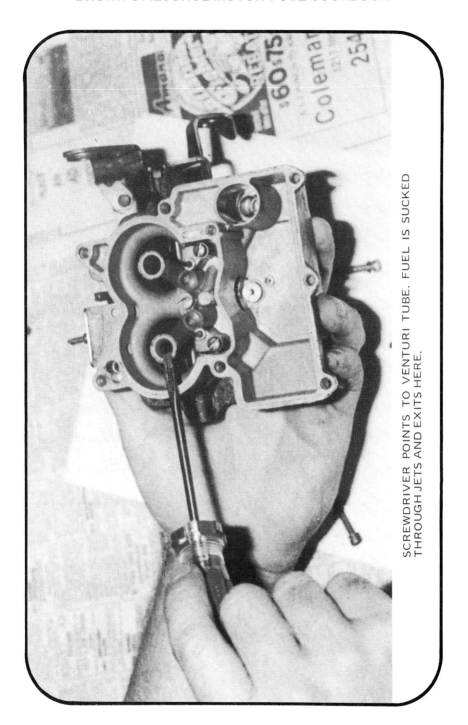

SCREWDRIVER POINTS TO VENTURI TUBE. FUEL IS SUCKED THROUGH JETS AND EXITS HERE.

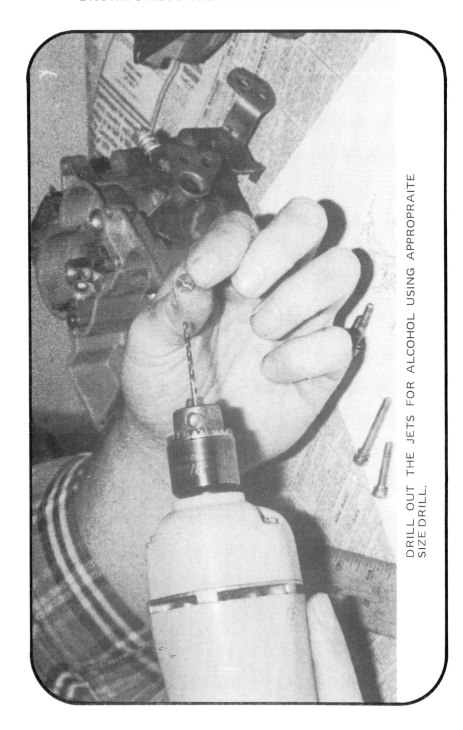

DRILL OUT THE JETS FOR ALCOHOL USING APPROPRAITE SIZE DRILL.

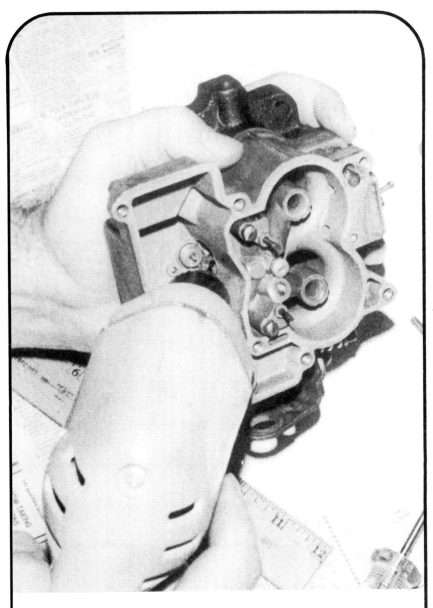

YOU CAN DRILL THEM OUT WHILE STILL IN THE CARBURE-
TOR. TRY NOT TO GET METAL SHAVINGS IN THE FUEL PAS-
SAGEWAYS. BLOW OUT WITH COMPRESSED AIR WHEN DONE.

weighs about 8 pounds per gallon. Alcohol weighs 6.6 pounds per gallon. Gasoline weighs 6.1 pounds per gallon. The weight of water is simply given as a reference point. Obviously alcohol that is mixed with water is going to weigh more than 6.6 pounds per gallon. Even at 6.6 pounds (anhydrous or pure 100% alcohol) alcohol is a thicker fluid than gasoline and will not flow as readily through a gasoline-sized jet.

However, the difference in the fuels by weight is only 10% or thereabouts. You would think a 10% enlargement of the main jet(s) would suffice. Sometimes it does but such a minimal enlargement does not work very well. Alcohol has only 2/3rds the energy, BTU's, calories, or whatever you want to call it that gasoline does. Add the 1/3rd for energy loss and the 10% for difference in fuel weight and what you get is a figure pretty close to the 40% range. All measurements given here are approximate since this is not a book for machine shop students or mathematicians.

The easiest way I have ever found to calculate the size drill bit I need for a main jet is to simply use a machinists tap and drill chart for decimal equivalents. If my two-barrel Rochester for my 1969 Chevrolet has a main jet size of .059 ("fifty-nine thousandths"), I look on the chart and find the number closest to .059. What I find is .0595 which is close enough. I look to the left of .0595 and find that is a No. 53 drill size. I'm not going to use that size drill, I just want to make sure I know what I'm doing. I then take .059 and multiply it by 40%.

What I get is .0236.

I add .059 and .0236.

Rounding the figures out to .060 x 4 equals .024 plus .060 equals .084 (that is, an .084 sized jet) is not as precise but will work if you're in a hurry (or hate arithmetic).

Doing it to the fourth decimal place I get .0826. I then look at my tap and drill size and there is no .0826 in the decimal equivalent column. However, there is a drill that matches .082. No. 45. To modify my carburetor to handle alcohol all I have to do is bore out my main jets with a No. 45 drill bit chucked into my handy Black & Decker. Had I used the rougher figure I would have wound up with either a No. 45 or No. 44 drill, either of which would have worked.

DECIMAL EQUIVALENTS AND TAP DRILL SIZES

FRACTION OR DRILL SIZE		DECIMAL EQUIVALENT	TAP SIZE	FRACTION OR DRILL SIZE		DECIMAL EQUIVALENT	TAP SIZE
					39	.0995	
	NUMBER SIZE DRILLS 80	.0135			38	.1015	5 - 40
	79	.0145			37	.1040	5 - 44
$\frac{1}{64}$		.0156			36	.1065	6 - 32
	78	.0160		$\frac{7}{64}$		.1094	
	77	.0180			35	.1100	
	76	.0200			34	.1110	
	75	.0210			33	.1130	6 - 40
	74	.0225			32	.1160	
	73	.0240			31	.1200	
	72	.0250		$\frac{1}{8}$		.1250	
	71	.0260			30	.1285	
	70	.0280			29	.1360	8 32,36
	69	.0292			28	.1405	
$\frac{1}{32}$	68	.0310		$\frac{9}{64}$		.1406	
		.0312			27	.1440	
	67	.0320			26	.1470	
	66	.0330			25	.1495	10 - 24
	65	.0350			24	.1520	
	64	.0360			23	.1540	
	63	.0370		$\frac{5}{32}$		.1562	
	62	.0380			22	.1570	
	61	.0390			21	.1590	10 - 32
	60	.0400			20	.1610	
	59	.0410			19	.1660	
	58	.0420			18	.1695	
	57	.0430		$\frac{11}{64}$		.1719	
$\frac{3}{64}$	56	.0465			17	.1730	
		.0469	0 - 80		16	.1770	12 - 24
	55	.0520			15	.1800	
	54	.0550			14	.1820	12 - 28
	53	.0595	1 - 64, 72		13	.1850	
$\frac{1}{16}$		.0625		$\frac{3}{16}$		.1875	
	52	.0635			12	.1890	
	51	.0670			11	.1910	
	50	.0700	2 - 56, 64		10	.1935	
	49	.0730			9	.1960	
	48	.0760			8	.1990	
$\frac{5}{64}$		.0781			7	.2010	$\frac{1}{4}$ - 20
	47	.0785	3 - 48	$\frac{13}{64}$		.2031	
	46	.0810			6	.2040	
	45	.0820	3 - 56		5	.2055	
	44	.0860			4	.2090	
	43	.0890	4 - 40		3	.2130	$\frac{1}{4}$ - 28
	42	.0935	4 - 48	$\frac{7}{32}$		.2188	
$\frac{3}{32}$		.0938			2	.2210	
	41	.0960		LETTER SIZE DRILLS 1		.2280	
	40	.0980		A		.2340	

DECIMAL EQUIVALENTS AND TAP DRILL SIZES

FRACTION OR DRILL SIZE	DECIMAL EQUIVALENT	TAP SIZE		FRACTION OR DRILL SIZE	DECIMAL EQUIVALENT	TAP SIZE
15/64	.2344			19/32	.5938	
B (LETTER SIZE DRILLS)	.2380			39/64	.6094	5/8
C	.2420			5/8	.6250	
D	.2460			41/64	.6406	
1/4 E	.2500			21/32	.6562	3/4 10
F	.2570	5/16 18		43/64	.6719	
G	.2610			11/16	.6875	3/4 16
17/64	.2656			45/64	.7031	
H	.2660			23/32	.7188	
I	.2720	5/16 24		47/64	.7344	
J	.2770			3/4	.7500	
K	.2810			49/64	.7656	7/8 9
9/32	.2812			25/32	.7812	
L	.2900			51/64	.7969	
M	.2950			13/16	.8125	7/8 14
19/64	.2969			53/64	.8281	
N	.3020			27/32	.8438	
5/16	.3125	3/8 16		55/64	.8594	
O	.3160			7/8	.8750	1 8
P	.3230			57/64	.8906	
21/64	.3281			29/32	.9062	
Q	.3320	3/8 24		59/64	.9219	
R	.3390			15/16	.9375	1 12
11/32	.3438			61/64	.9531	
S	.3480			31/32	.9688	
T	.3580			63/64	.9844	1 1/8 7
23/64	.3594			1	1.0000	
U	.3680	7/16 14		1 3/64	1.0469	1 1/8 12
3/8	.3750			1 7/64	1.1094	1 1/4 7
V	.3770			1 1/8	1.1250	
W	.3860			1 11/64	1.1719	1 1/4 12
25/64	.3906	7/16 20		1 7/32	1.2188	1 3/8 6
X	.3970			1 1/4	1.2500	
Y	.4040			1 19/64	1.2969	1 3/8 12
13/32	.4062			1 11/32	1.3438	1 1/2 6
Z	.4130			1 3/8	1.3750	
27/64	.4219	1/2 13		1 27/64	1.4219	1 1/2 12
7/16	.4375			1 1/2	1.5000	
29/64	.4531	1/2 20				
15/32	.4688					
31/64	.4844	9/16 12				
1/2	.5000					
33/64	.5156	9/16 18				
17/32	.5312	5/8 11				
35/64	.5469					
9/16	.5625					
37/64	.5781	5/8 18				

PIPE THREAD SIZES

THREAD	DRILL	THREAD	DRILL
1/8–27	R	1 1/2–11 1/2	1 47/64
1/4–18	7/16	2–11 1/2	2 7/32
3/8–18	37/64	2 1/2–8	2 5/8
1/2–14	23/32	3–8	3 1/4
3/4–14	59/64	3 1/2–8	3 3/4
1–11 1/2	1 5/32	4–8	4 1/4
1 1/4–11 1/2	1 1/2		

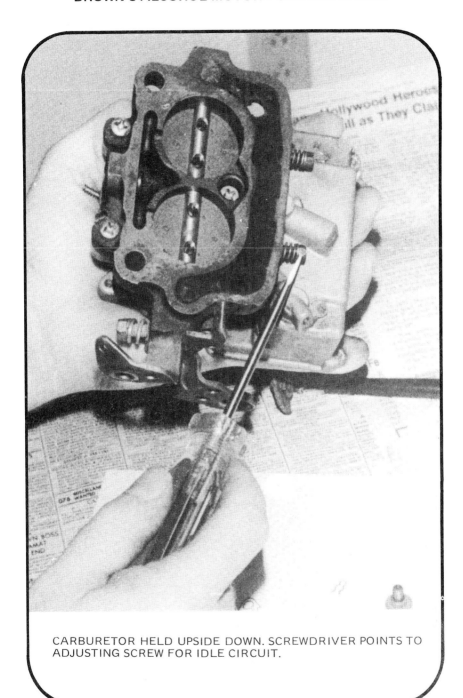

CARBURETOR HELD UPSIDE DOWN. SCREWDRIVER POINTS TO ADJUSTING SCREW FOR IDLE CIRCUIT.

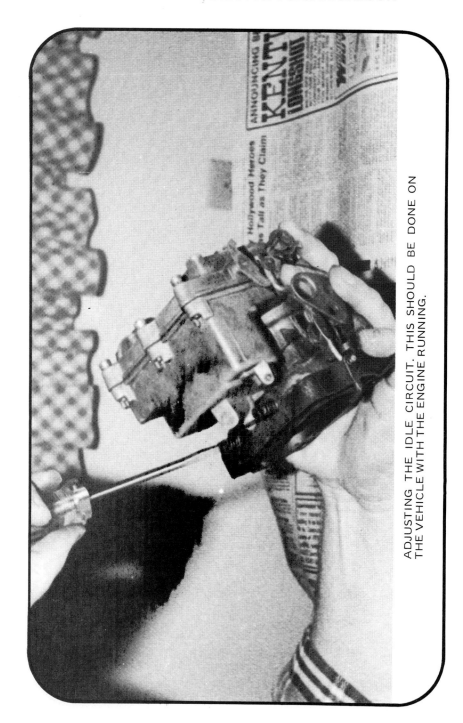

ADJUSTING THE IDLE CIRCUIT. THIS SHOULD BE DONE ON THE VEHICLE WITH THE ENGINE RUNNING.

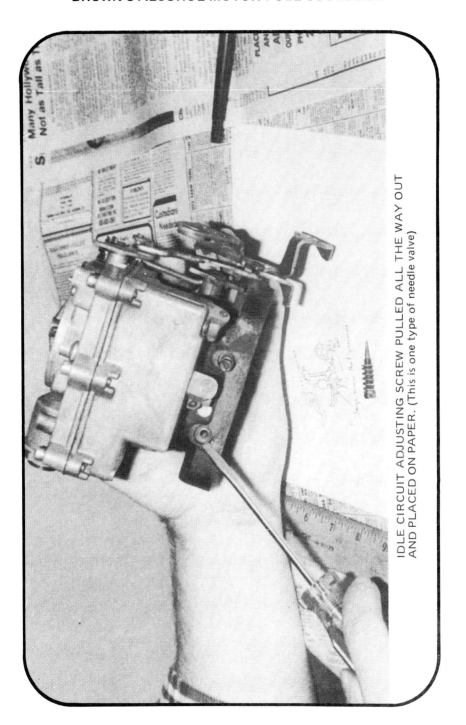

IDLE CIRCUIT ADJUSTING SCREW PULLED ALL THE WAY OUT AND PLACED ON PAPER. (This is one type of needle valve)

The number No. 44 would cause the engine to consume more fuel, though.

At this point it is time for you, me, or us to reassemble the carburetor, stick it back on the vehicle, and run our first road test.

Be prepared for a couple of disappointments.

The first is that our car won't idle.

We didn't set the screws on the bottom of the carburetor, which control what mechanics call the idle circuit, for alcohol. They have to be opened up. At this point if we can feather the accelerator pedal and keep it running; let's just do that. We'll cover needle valves and such a little further on.

The next thing we notice is that our car doesn't seem to have any oomph at low speed (or rpm's). Most race car drivers never notice this since almost all their driving is done with their right foot all the way to the floor. Alcohol burns slower than gasoline and this is where we find out about it. It is especially noticeable on low-compression engines. On my old 1948 74 Harley-Davidson with 6.6 to 1 compression ratio it was all I could do to get it going more than 20 mph up a steep hill in the neighborhood. In most cases an old Harley-Davidson will literally walk up the side of a building. I could modify the machine to run on alcohol where it would out-climb and out-run any other comparable Harley in Kentucky if I was willing to forfeit gasoline completely as a motor fuel.

On the plus side, the "steam engine" effect of alcohol in a gas engine makes for an incredibly smooth ride down the road. The slower burning alcohol appears to lessen engine vibration considerably.

The last thing we notice as we refill the fuel tank for another road test is that we used a bit more fuel than we did with gasoline. If we got 18 mpg with gasoline all we got with alcohol was 16 mpg. This is, of course, if all we did was drill out the main jet or jets.

Fuel economy and power have been measured by a number of research departments using what are known as "bench tests." A stationary engine is hooked up to a dynamometer, emission control equipment, and so on. I did the same thing at Berea College. Such tests leave much to be desired since you can't climb a hill with an engine bolted to a shop floor and the air flow into the carburetor air horn simply isn't the same. There are too many variables. Not to mention most bench tests are almost always conducted at full load and full throttle.

If you drill your jets too large or too small a number of undesirable consequences occur.

If your drill is too large the resulting hole causes enormous fuel consumption. Alcohol will keep burning in an engine long after the same percentage (proportionately) of gasoline would have simply flooded and stalled the engine. In one case I heard of a race-car driver had his main jets removed by his mechanic without his knowledge and drove off anyway: shooting 30 foot flames out of his tailpipes. Since alcohol has a generous supply of oxygen in the compound its propensity to keep running might be explained.

If your jet size is too small you are subject to burn your valves. A gasoline-fueled engine when the jets are too small will sputter and misfire. An alcohol-fueled engine with slightly undersized jets will simply burn a lot hotter, which in turn burns valves. Most American cars are designed to use unburned fuel and tetraethyl lead to lubricate the exhaust valves. This might not be a problem with engines designed to use unleaded gas since they come from the factory with hardened valve seats. Of course, fuel use or consumption goes down with smaller jets.

Valve burning can be prevented several ways. One is to simply chuck a half-cup of vegetable oil into the fuel tank with the alcohol. Diesel fuel (a half-cup) could be used in a pinch but is not recommended since alcohol and petroleum products will not mix if there is any water in the alcohol (there usually is).

If the engine runs cool enough water in the alcohol will sometimes act as a valve lubricant. Sometimes.

The best procedure to follow if you are really serious is to install stellite or stainless steel racing valves with hardened seats. A lot of propane-powered vehicles are modified in this fashion since the propane burns completely dry in an engine.

If all you intend to do is go out and buy a spare carburetor for the day the gas pumps shut down there are a few pieces of information you might want to digest first.

You can normally put a Ford carburetor on a Chevrolet and vice versa with the aid of adaptor plates. You might want to consider it if push comes to shove since some carburetors are a whole lot easier to work on than others. A two-barrel Rochester has to be completely taken apart to get at the jets. You don't have to take the guts out of it but you do have to disconnect all the linkage and separate the top and bottom halves. A Holley has the jets behind the float and all that is required is for the float cover to be unbolted and the jets screwed in or out. Holleys have horizontal jets. That is, the holes in them are parellel to the horizon when the carburetor is mounted on the vehicle. Almost all others have vertical jets.

If you are willing to go to the extra trouble and expense you don't have to get out the toolbox every time you want to switch from gasoline to alcohol. Or back. There is only one thing required.

It's called a needle valve and at one time almost all carburetors came equipped with them. A tapered shaft was inserted into the main jet by means of a screw thread adjustment. The screw thread adjustment allowed the tapered shaft to be moved in and out of the hole in the jet. The further the shaft was inserted into the hole the smaller the opening (the hole) became. As it was unscrewed or withdrawn the opening became larger, which meant that if the hole area was .059 all it would take is a flick of the wrist to bring it up to .082.

Since such precise measurements are almost impossible with the wrist, either the setting had to be done by ear or an rpm gauge had to be used; or a vacuum gauge. Ecology fans sometimes use emission control gauges (like on a Sun Tester) to set the opening, which is not a bad idea since less unburned hydrocarbons means that more fuel is being utilized by the engine which in turn means better mileage.

Most American car builders went to the fixed-jet principle with the advent of the down-draft carburetor. Most of the old updrafts (the air was sucked from the bottom up

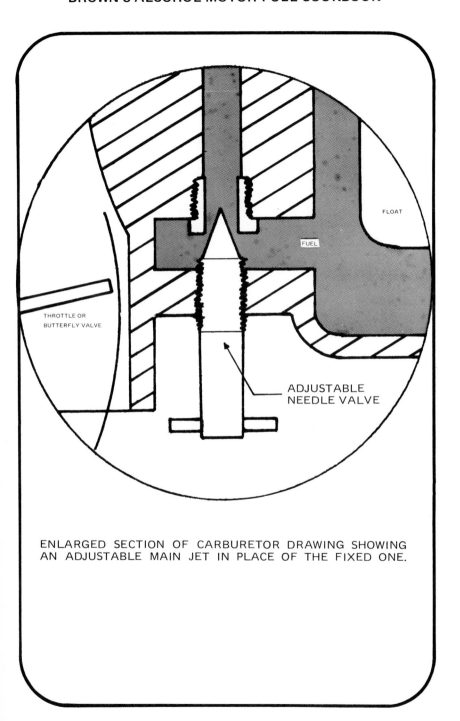

ENLARGED SECTION OF CARBURETOR DRAWING SHOWING AN ADJUSTABLE MAIN JET IN PLACE OF THE FIXED ONE.

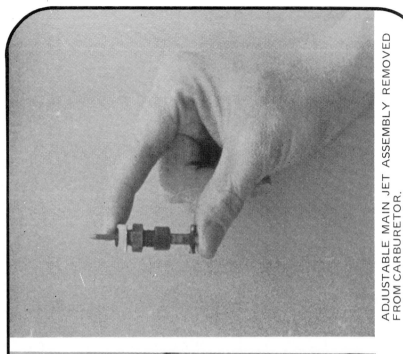

ADJUSTABLE MAIN JET ASSEMBLY REMOVED FROM CARBURETOR.

CARBURETOR WITH CUT-A-WAY SECTION TO SHOW THE ADDITION OF AN ADJUSTABLE MAIN JET.

instead of the top down) had adjustable needle valves. Main jet adjustments come in quite handy for altitude adjustments as anyone who has ever driven from L.A. to Denver will tell you.

If you don't want to mess with all the rpm and vacuum gauges, you don't really have to. It's not mandatory. All that is required is a good "ear" to listen to the engine with. Start unscrewing the needle valve outward when you are going from gasoline to alcohol. On too lean a mixture of alcohol the engine will hesitate and sputter. Simply listen for the engine "smoothing out" and that's the setting you run it on. It might take a little practice. If you have ever used a power lawnmower that had a carburetor adjustment then you have already had the practice.

There are a few engines and a carburetor or two still on the market that come factory-equipped with a main jet adjustment. Facet carburetor corporation markets a number of carburetors with main jet adjustments for stationary power plants. Their primary drawback for automotive use is that such carburetors have no accelerator pump, meaning you might be able to get from point A to point B if you don't mind creeping through every intersection, never passing another car, and having your car die on you every time you push the gas pedal down a little too far.

Many small air-cooled engines have main jet adjustments. Normally labeled as needle valves in the parts manuals. I have a 3 hp Briggs & Stratton with a Flo-jet carburetor that has one that I use for demonstrations — just twist the needle valve adjustment out one and a half turns, pour in a pint of liquor store whiskey, and fire it up.

My ten year old son uses rubbing alcohol in two of his Tecumseh engined mini-bikes. Same adjustment. He doesn't have to impress anybody with or by breaking a government seal and the rubbing alcohol is a whole lot cheaper. Obviously rubbing alcohol is a lot more expensive than gasoline, but it still beats walking.

The only full-sized vehicle still equipped with a main jet adjustment when it comes from the factory is probably the Harley-Davidson. At least, that's what the catalogs and

MOST AMERICAN MADE SMALL AIR-COOLED ENGINES HAVE AN ADJUSTABLE NEEDLE VALVE ALREADY PRESENT IN THE CARBURATOR. ENGINE SHOWN ABOVE IS A TECUMSEH 3hp.

parts manuals indicate. On my old 1948 model there are two needle valves for the Linkert carburetor, both for high and low speed.

If you look at the base of most passenger car carburetors you will see a screw, sometimes with a spring around it (two on a two-barrel). That is a needle valve controlling the idle adjustment. It is possible to back that screw(s) out and get a standard carburetor to idle and nothing else on alcohol. That adjustment was left on since at idle an engine is normally running on a 6 to 1 air to gasoline mixture instead of the normal 15 to 1. The percentages are far more critical.

In order to avoid the constant hassle of pulling carburetors apart to change fuels and since I couldn't find a suitable carburetor with a main jet adjustment, I had an adjustable needle valve cabbaged from a Briggs & Stratton installed in a one-barrel Rochester. Some of the base had to be milled or ground away, the location of the main jet inside the float cover was determined, a hole was drilled and tapped, the needle valve was turned down to fit, and the main jet drilled out to .094. The .094 is 60% oversize since that also allows wood alcohol to be used (which has 73% of the BTU's of alcohol from starch crops, sugar, or fruit). It's not hard if you can center the drilled and tapped hole properly. It's impossible if you can't.

If you have some skill as a machinist a Quadra-jet carburetor can be reworked a number of different ways, including some that I'm sure I'll never discover until someone else points them out to me.

The back two barrels on almost any four-barrel carburetor such as the Rochester are almost always used only for passing and the like. Normally no gasoline will flow through the back two jets. On a Rochester the back two jets are opened and closed by two metering rods that function almost the same way a needle valve does. Tromp on the accelerator pedal and the metering rods rise up out of the jets, which should mean that if the main jets are blocked off in the front two barrels (block off the entire section and you would have no idle setting), the metering rods remachined to a taper, the jets bored out, and the linkage reset (unless you like driving around at 40 mph with your foot all the way to the floorboard) you should be able to run either gasoline or alcohol depending on how far up or down the rods are in the jets.

I say should, because if you aren't an experienced mechanic or machinist don't even think about it.

Another method is to simply buy two Rochesters, cut off the back two barrels with the metering rods in them on both of them, set them aside or throw them away, turn one around backwards and weld them together. Welding pot metal isn't easy but there are folks who can do it. Install a gasoline fuel line to either the front or back half of the carburetor. Enlarge the jets and install an alcohol fuel line to the other half of the carburetor. Run the alcohol and gasoline fuel lines to their respective fuel tanks with shut off valves. You now have a dual-fuel carburetor. Obviously you are going to have to fabricate an adaptor plate for the intake manifold.

There was a carburetor manufactured up until 1959 by a man named John Robert Fish that operated on a self-adjusting principle right from the factory. That is, with no manual manipulation of anything, it would run well on alcohol, gasoline, or kerosene. A number of people around the country are trying to get it back into production since it normally gave 40% better gas mileage.

What Mr. Fish did was drill his jets (six in all) in the butterfly (or throttle) valve of a one-barrel carburetor. The fuel passage from the jets made a right-angle turn down a radial arms that swung at the bottom of the fuel bowl and picked up fuel from a groove cut in the side of the float chamber. The groove was deeper and wider at one end than it was at the other. The further the throttle was opened the deeper and wider the groove became under the hole in the radical arm, in effect increasing the size of the main jet larger and larger as the gas pedal was pushed further and further down, which meant that the accelerator pedal simply had to be pushed closer to the floorboard when the fuel was alcohol than it did for gasoline.

There is one other carburetor adjustment possible if what you are after is maximum economy. As anyone who has

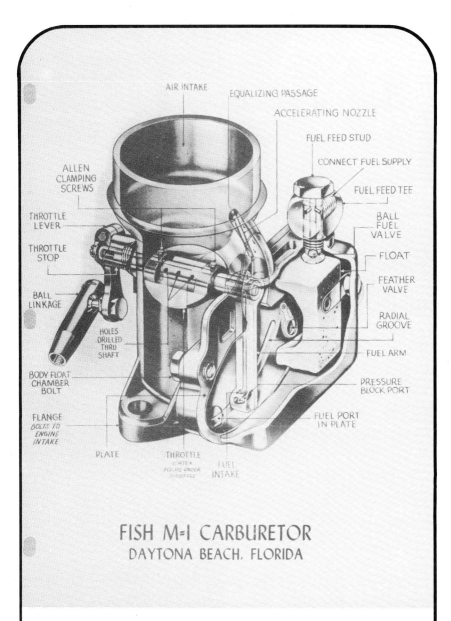

AIR INTAKE

EQUALIZING PASSAGE

ACCELERATING NOZZLE

FUEL FEED STUD

CONNECT FUEL SUPPLY

ALLEN
CLAMPING
SCREWS

FUEL FEED TEE

THROTTLE
LEVER

BALL
FUEL
VALVE

THROTTLE
STOP

FLOAT

BALL
LINKAGE

FEATHER
VALVE

RADIAL
GROOVE

HOLES
DRILLED
THRU
SHAFT

FUEL ARM

BODY FLOAT
CHAMBER
BOLT

PRESSURE
BLOCK PORT

FLANGE
BOLTS TO
ENGINE
INTAKE

FUEL PORT
IN PLATE

PLATE

THROTTLE
WHEN
FOLDS UNDER
THROTTLE

FUEL
INTAKE

FISH M=1 CARBURETOR
DAYTONA BEACH, FLORIDA

THE PHOTOGRAPH ABOVE IS TAKEN FROM A SALES BRO—
CHURE ON THE FISH M-1 CARBURETOR. AN INTERESTING
DEPARTURE FROM "STANDARD" DESIGN, THE FISH M-1 WAS
AVAILABLE DURING THE EARLY 1950s.

worked on carburetors knows, float adjustment can alter fuel economy. Rather than go through a lot of trial and error adjustments the easiest thing to do is simply weigh the float on a gram scale, figure out what 10% of the total weight of the float is, and braze (or glue if you can get any to hold) that amount of weight to the top middle of the float. Remember, alcohol as a fluid is 10% heavier than gasoline.

The 100 Miles per Gallon

Alcohol Carburetor

You have probably heard all the stories of someone's uncle who bought a new car, drove it around for awhile, and found he was getting 50 miles to the gallon of gasoline. You may have pooh-poohed the stories.

Don't.

There have been numerous inventors who tried to build such carburetors and succeeded. Every single one of them was designed and built on easily verifiable physical laws. Starting with the first one in 1912, including several built by Charles Pogue of Winnipeg, Canada in the 1930's that consistently achieved 200 miles per gallon in V-8 Fords, and recently reappeared in a 100 mile per gallon device built by a mechanic named Tom Ogle from Texas.

The standard carburetor is a rather crude device compared to what it could be. Fuel injection isn't much better. What happens when the fuel is sucked into the engine explains why today's carburetors are so crude.

As the gasoline passes from the venturi tube of the carburetor into the intake manifold and then into the cylinder it is supposed to be turned into a vapor. Only a vapor will ignite under the spark plug, small liquid droplets either burn or are simply heated up enough to leave in the exhaust manifold in a vapor state (too late to get the job done). A standard carburetor and intake manifold vaporize part of the fuel and allow part of it in the form of small liquid droplets into the combustion chamber. If it gets too bad in the intake manifold it is known as "puddling" or flooding. The compression stroke raises the internal temperature of the cylinder high enough to vaporize more fuel but still leaves it far short of the complete vapor state which is how and why the 50 mile per gallon carburetors work. The gasoline is simply turned into a complete vapor before it is allowed to enter the combustion chamber or cylinder which in turn means a lot more energy is extracted from it. You can verify this in a high school chemistry book. The auto manuals tell us the correct mixture for a standard carburetor is 15 pounds of air to 1 pound of fuel. Yet a high school chemistry book tells us that the correct explosive mixture for a vapor — any vapor — is 50 pounds of air to 1 pound of explosive liquid. Apparently two-thirds of our fuel is being wasted and we are only counting the power in the hydrogen of an explosive hydrocarbon.

Another way you can verify this tremendous fuel loss (or inefficiency) in our modern engines is to simply hook a spark plug into your tailpipe and wire it up to your electrical system. Make sure you have a switch or other method of disconnecting it. Drive down the road at night and flip it on. The unburned fuel coming out of the tailpipe will cause 30 to 40 foot flames.

The high-mileage carburetors did have a number of problems. The first obstacle that had to be overcome was the complete vaporization temperature of gasoline: well over 400 degrees F. Some parts of the gasoline, ordinary gasoline is actually a series of compounds, vaporize as low as 90 degrees F., but only some, and 400 degrees heat causes other problems. Like warming the incoming air so much that the engine loses considerable power.

The main obstacle to any type of production was the amount of plumbing or gadgetry required to make a vapor-phase carburetor work. The ones built by Charles Nelson

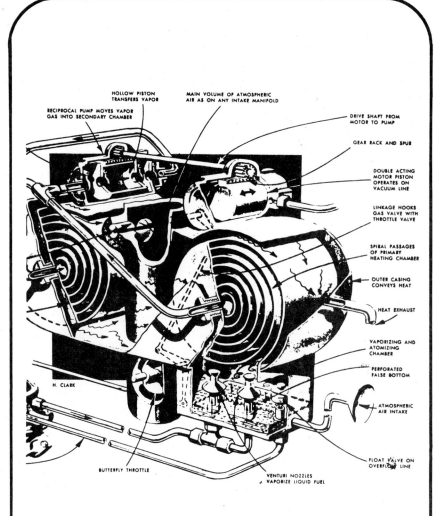

A 200 MPG GASOLINE CARBURETOR
These were built in the 1930s by Charles Nelson Pogue.

Reprinted by permission of Fawcett Publications

Pogue were almost the size of the engine itself, and a lot more complicated.

It is the opinion of this writer that the Pogues, the Ogles, and all the others had the right idea. They just used the wrong fuel.

Had they used alcohol, the attempted (and sometimes successful) attainment of 400 plus degree temperatures would have been totally unnecessary. Alcohol is a single compound that vaporizes at a temperature of less than 180 degrees F. Over 30 degrees F. less than the boiling point of water. Plus all the parts for such a carburetor are already production items available almost everywhere in the United States.

Propane parts will do for alcohol what $100,000 worth of tooling and research has done for gasoline. The reasoning and arithmetic is astoundingly simple.

Propane and gasoline will give almost the same mileage in the same vehicle. In many cases a car getting 20 miles per gallon will deliver 18 miles per gallon on propane. Yet the energy available in a gallon of gas is enough to lift a 3,000 lb. car 7 miles straight up in the air. Logically and proportionately the gasoline should move the vehicle eight times the distance that it does. In the case of Charles Nelson Pogue, it did.

Propane vaporizes at 30 degrees F. below zero. By the time it gets to the engine block it is a complete vapor. The way gasoline should be. The low vaporizing point of propane is the reason for keeping it in pressurized tanks. In some cases propane is passed through a pressure regulator filled with hot water to vaporize the fuel. Such hot water, if above 180 degrees F. and the passage fuel goes through is long enough to raise it to the same temperature, will also completely vaporize alcohol. A propane carburetor must be used since a vapor will not work in a normal carburetor with jets and float bowl.

This gives us the following mileage figures: if gasoline has eight times the energy utilized and alcohol has two-thirds the energy of gasoline then alcohol should have roughly five and one-half times the energy available. If our vehicle runs

18 miles per gallon on gasoline and we multiply that performance by five and one-half we arrive at a figure of 99 miles per gallon. If, of course, the alcohol is completely vaporized. It normally is if the cooling water on a condensor on an alcohol still is allowed to get too hot. The same principle applies to a motor vehicle engine.

At this point I realize there are a lot of skeptics. Don't knock it until you have tried it. Charles Kettering, inventor of the electric starter and one of the "big guns" of the automotive world, said, "If the compression ratio of an engine of 6.5 to 1 were raised to 10 or 12 to 1, little gain in power and efficiency should be expected due to the internal friction which is brought about by their lack of rigidy. Roughness, increased friction and other mechanical problems tend to counteract any gains from high compression ratios." He said this in 1947, over thirty years after he had invented the electrical starter. Published in MORE EFFICIENT UTILIZATION OF FUELS for the Society of Automotive Engineers summer meeting of June 1 - 6, 1947. Thirty years after his remarks, Detroit was spewing out engines with 10.5 to 1 compression ratios by the trainload and some engine builders, such as Chrysler, sell off-the shelf parts to enable a car owner to raise his compression ratio to 12 to 1. Just because someone says something, including an expert, does not make it so. If Charles Kettering can be so wrong so can the skeptic down the street.

There is one thing to watch out for — in some systems the alcohol recondenses going down the venturi tube and a cancellation effect takes place.

Compression

How much the piston compresses the fuel in the combustion chamber determines to a large extent how much energy is extracted. Back in 1906 folks at the Department of Agriculture thought they had really done something when they ran alcohol-fueled engines on compression pressures of 180 lbs per square inch. Tetraethyl lead hadn't been discovered and gas at that pressure would simply blow the engine apart. Today most gasoline powered engines operate in that pressure range.

The compression ratio of most full-size auto engines manufactured in the last fifteen to twenty years is usually 8.5 to 1 or better. If you run alcohol in your engine you won't have anywhere near the power loss problems I had in my old Harley-Davidson. However, to extract maximum power and economy from alcohol, the compression ratio can be raised to 15 to 1 on most engines. If you intend to use gasoline also you will have to compromise since at 15 to 1 gasoline will simply blow the engine apart. 11.5 to 1 will work if you stick strictly to premium gasoline. There are a number of ways to modify the engine in this regard.

The most expensive ways appear to be the most well-known. Install high compression pistons. Pull the cylinder heads off, strap them to a milling machine, and remove some of the metal, called "milling the heads." Take the main body of the engine and shave more metal from it, called "decking the block." Expensive, complicated, and time consuming.

The easiest way to raise a compression ratio on a large engine is to simply pull off the stock cylinder heads and replace them with ones from a smaller engine. For example, if you have a 350 engine just go to a junkyard and pick up a set of 265 heads. Be sure your junkman looks in his parts book and verifies that the 265 heads from that make and model have the same bolt pattern as yours and will bolt right on. Normally this will raise your compression ratio to the 11.5 to 1 range. If you have a large-block engine, you may have a problem.

It is not even necessary to clean the carbon off the heads if you are going to run alcohol first. Just bolt the nasty things on. The alcohol will clean them up as you drive. Before World War II many Europeans would run a tank of alcohol through their cars every now and then because of the natural biological solvent properties of alcohol. The alcohol would pick up all the gum and varnish out of the fuel tank, fuel lines, and carburetor (sometimes plugging up the jets), and clean all the carbon off the cylinder heads. If you have carbon deposits on your present cylinder heads it's easy to tell. Your car will keep running after the ignition has been shut off since the carbon desposits become hot enough to ignite the fuel on their own, called "dieseling."

How fast the alcohol cleans the carbon off is amazing and has to be seem to be believed. The head chemist of the Alcohol Fuels Corporation of Gering, Nebraska and I were running alcohol through a variable-compression engine one day to see how high a compression ratio we could attain. He would raise the compression ratio and the engine would start knocking. In a few seconds a layer of carbon would be cleaned off and he would raise the compression again. The engine would start knocking again, the next layer of carbon would be cleaned off, he would raise the compression again, and so on.

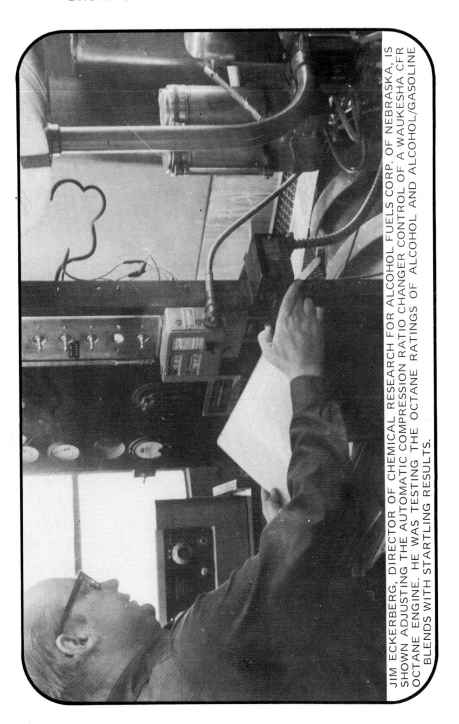

JIM ECKERBERG, DIRECTOR OF CHEMICAL RESEARCH FOR ALCOHOL FUELS CORP. OF NEBRASKA, IS SHOWN ADJUSTING THE AUTOMATIC COMPRESSION RATIO CHANGER CONTROL OF A WAUKESHA CFR OCTANE ENGINE. HE WAS TESTING THE OCTANE RATINGS OF ALCOHOL AND ALCOHOL/GASOLINE BLENDS WITH STARTLING RESULTS.

Ignition

Alcohol is a cooler and slower burning fuel than gasoline. The slower burning requires an advanced ignition timing. That is, the spark plug must fire at a point or time before what it would require for gasoline. Timing for alcohol is not a mandatory procedure but it does help to give better fuel economy and power.

Timing for gasoline is usually set by a timing mark on a flywheel pulley in front of the engine and a strobe light. The timing mark is put on at the factory and it is not always accurate. If it is, then for alcohol you must set the timing somewhere ahead of it. The accepted procedure is to simply time it on gasoline, start it up on alcohol, and then turn the distributor until the engine starts knocking, a few degrees in front of the position it begins to knock is your setting. If you want to be really scientific about it use a stethoscope to pick up all the engine noises and knocks.

The absolute best way to go back and forth from gasoline to alcohol is to install a spark advance on the column, like the old model A Fords had. Once installed you can change timing in an instant.

If you are going to run alcohol only in your engine, you may want to run hotter spark plugs in your engine. You will have to experiment.

Cold Weather Starting

After several weeks of running bench tests at Berea College with my alcohol-fueled Briggs & Stratton, I thought I had the energy crunch solved. Alcohol polluted less. It gave more power at high rpm. It could be produced from renewable starch and sugar crops. It started on the first crank every morning, better than gasoline.

Or so I thought.

On a cold morning in November I walked into the shop, set the choke on the carburetor as its usual setting, and yanked on the recoil starter. It should have started.

But it didn't.

Half an hour later I was still yanking on the starter cord and nothing was happening. It finally dawned on me that something was wrong. Either something with the engine or the fuel. Since the fuel was the easiest to check I drained the tank and refilled it with gasoline. It started right up. Therefore, it had to be the fuel, which led me to the conclusion that alcohol would not ignite readily in freezing weather.

I was only partially correct.

A few minutes after I got the cold engine running on gasoline I switched back to alcohol just to see what would happen. It worked perfectly. My next conclusion was obviously that in cold weather an engine had to be started on gasoline first. All I had to do was get the engine warm.

A two-way directional valve would solve running the different fuels into the same carburetor but then there would be the problem of different jet sizes. A dual-fuel Rochester or a carburetor with an adjustable needle valve would work. The other alternatives simply consist of a spray mist of gasoline introduced into the engine via a nasal spray bottle (just hold it over the carburetor air horn and squeeze, hook up a windshield spray device but load it with gasoline, or use a diesel starter fluid tube). Don't use a bottle of ether unless you know exactly what you are doing. The stuff has a tendency to blow head gaskets.

Chrysler solved the problem in 1936 in cars exported to New Zealand that were set up to run alcohol from the factory (New Zealand had no domestic oil supplies) by putting heat on the intake manifold under the float bowl. Once an engine is started it will run on alcohol at 20 degrees F. below zero with no problem.

A method used early in the 20th century was to simply add benzine (from petroleum) or benzene (from coal tar) to the alcohol to raise the vapor pressure. Messy.

The best method I have come across is to simply install a small sealed fuel tank off to one side of the engine with a heater element immersed in the fuel. A tube leads from the small fuel tank to the air cleaner. When the heater element is activated the alcohol heats up past 180 degrees in a matter of a minute or two, the vapors rise into the air cleaner, and the vaporized fuel ignites instantly when the engine is cranked over and the fuel makes contact with the spark plug flame.

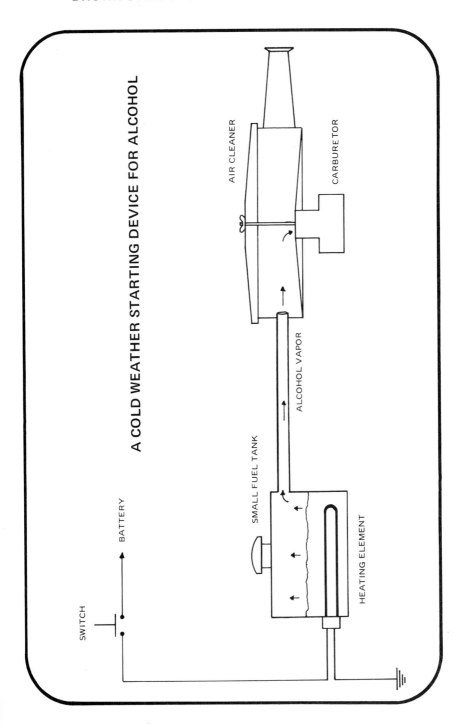

A COLD WEATHER STARTING DEVICE FOR ALCOHOL

AIR CLEANER

CARBURETOR

ALCOHOL VAPOR

SMALL FUEL TANK

HEATING ELEMENT

BATTERY

SWITCH

The Basics of Producing

Alcohol for Fuel

Ethyl alcohol for fuel use has to be stronger than that used for drinking. If it won't burn when it is in a bowl on the table and you have struck a match to it neither will it burn in your car engine.

To produce such alcohol, you start with either sugar or a starch. If you use a sugar, such as rotten fruit from the local grocery store, you simply ferment it until it smells, add a pinch of yeast, and distill the product. The yeast consumes the sugars and excretes alcohol and carbon dioxide.

Just letting the yeast act on the sugars will never raise out concentration of alcohol high enough to power an engine. All if does is make wine at this stage.

To increase the percentage of alcohol in the fluid we have to distill it. Distillation consists of purifying the liquid by boiling it, collecting the vapor, and changing it back to a liquid again by passing over a cold surface, leaving the impurities behind.

That is all there is to it for simple sugars except that in the case of distillation Murphy's Law is almost always a major factor: if anything can go wrong it will. Be prepared for problems.

Starch crops are considerably more difficult. Starch is too long a carbon chain for the yeast to act on and must be converted to sugar through a catalytic reaction. The processes are called malting and mashing and will be covered later on in these pages.

A word needs to be said regarding alcohol "proof." Two alcohol proof equal one per cent. That is, 140 proof alcohol would consist of 70% alcohol and 30% water. 100 proof alcohol will work in an engine but it will not work very well. There is a tremendous power loss, over one-third of what the engine is capable of, and even harder to start than usual when cold. Oddly, there is no difference in power and mileage between 160 proof alcohol and 200 proof. At 160 proof the 20% water appears to cool the exhaust valves and at 200 proof all the moisture is sucked out of the engine. 200 proof alcohol will suck moisture right out of the air. Ever notice how much better your engine runs on a cool, damp night?

Do not be confused by anyone talking about "proof gallons." A proof gallon is 50% water and 50% alcohol (100 proof) by volume.

There are five processes and five temperatures required in distilling. Make one mistake and what you get is not alcohol: it will be either distilled water or mush.

The temperatures are:

212 degrees F.		boiling point of water
173	"	boiling point of alcohol
152	"	conversion of starch to sugar
90	"	yeast dies
62	"	yeast goes dormant

The five processes are malting, gristing (grinding), mashing, fermenting, and distilling. They are labeled on the flow chart and explained in the following chapters.

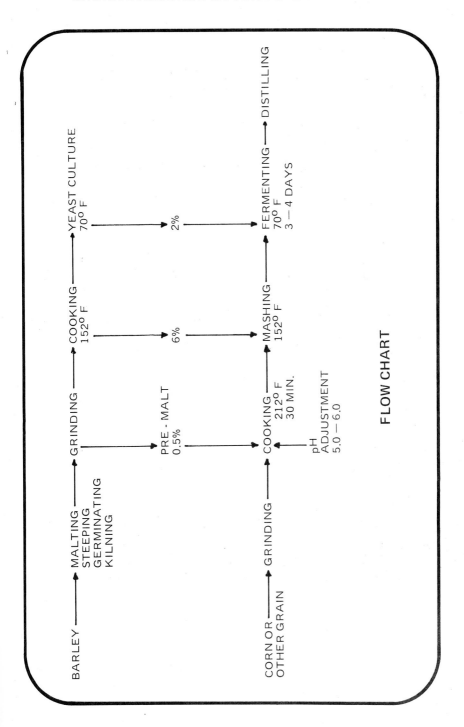

FLOW CHART

Malting

Malting is the basis or starting point for all distillation from starch crops. It is not necessary for fruit. Malting consists of converting the non-fermentable starch in grains such as corn, wheat, rice, and so on to sugar that is fermentable. Starch is a much longer carbon chain than sugar and is simply too strong for yeast to process.

Distillers' malt grain should be high in protein and low in starch. Brewers' malt grain should be just the opposite. Barley is normally malted since it is the easiest and fastest grain to work with but others, such as corn, can be used.

Your barley should have the following characteristics:

1. Even grain size
2. Fully ripe
3. Sweet and dry (good condition)
4. Low nitrogen content (indicated by being internally floury)
5. It should not have been treated with chemicals

Malting is divided into three phases:

1. Steeping (soaking)
2. Germinating (sprouting)
3. Kilning (drying in low temperature oven)

The process begins by putting a quantity of barley in a tub and covering it with six inches of water. It is soaked for 24 hours, in which period of time it should become stale and emit a foul odor. The water is drained and replaced with fresh water. This is let stand for at least another 24 hours, it can be soaked for up to a week without harm. Room temperature should be between 63 degrees and 86 degrees F. Do not let it get too cold.

The easiest arrangement for home production barley steeping is to place a bucket with a lot of small drain holes inside another, larger bucket with a drain cock or faucet. The water can be let out without up-ending the bucket full of barley and making a big mess. Not to mention that barley will soak up four times its own weight of water and the 56 lb bushel you put in the tub will weight 280 lbs when you try to get it back out.

That is all there is to steeping.

The next procedure is germinating or sprouting. It is basically the same procedure the health food folks do with their seeds.

The grain is kept moist until it begins to sprout. Not soaked, just moist. The easiest and most space-saving way is to simply spread the barley over fine wire mesh that the water can leak through. Spread it thin and sprinkle it enough to keep it moist. If you spread it more than an inch deep you will have to turn it twice a day to insure good ventilation.

If you don't want to mess with the screens and have a concrete floor handy you can use the 19th century method for volume production that didn't require any machinery.

Pile the grain up on the floor in 10 inch high heaps. The outside of the piles should become dry and the interior become warm. The barley is then mixed with great care to avoid breaking the seed. You should be mixing dry seed with wet seed and warm seed with cold seed. Repeat every six hours until the germ (or sprout) has grown as long as the seed.

It may take anywhere from 3 days to a week for the grain to sprout.

The germ is then removed with the aid of a coarse sieve.

Commercial outfits use scrubbers. Removing the sprouts is not absolutely necessary. Do not let the sprouts get longer than the seed.

The third procedure is kilning or drying. The sprouted barley is placed in a low temperature oven at slightly under 122 degrees F. until it appears to be dried completely. Sunlight will work on a hot, dry day. Pick it up with your fingers and feel it occassionally until it appears to be absolutely dry. When it feels totally dry, raise the oven temperature up to slightly under 140 degrees F. and let it bake some more. Anywhere from a half hour on up depending on how much local humidity you have. The more humid your air is the longer you need to cook it.

The result is malt: the basis of almost all ale, beer, porter, whiskey, bourbon, and home-made motor fuel alcohol. This should make a real hit with the health food crowd: organic motor fuel.

The malt is ground into coarse meal or crushed two days before use. Older malt often loses its secondary conversion activity which in turn results in a lower per bushel yield of alcohol.

Commercial distilleries normally use a hammermill, roller mill, or attrition mill to grind (or grist) the barley. You can do the same thing in your kitchen with a rolling pin or a coarse grinder. If you intend any sort of volume production, I suggest a coarse grinder with an electric motor.

The physical reasoning behind the malting process is as follows:

The enzymes in the grain are activated to convert starch to sugar. They multiply enormously to convert the starchy part of the grain to sugar that the young grain plant can utilize for food.

When the sprout is past 1/4 inch long the food supply starts to be used up.

At this point we terminate the process by introducing artificial drought conditions by heating the barley up to 120 degrees F. The sprout dies and withers away but the enzymes remain in the seed full of life.

The sprouted barley is then run through our coarse grinder to allow enzymes and grains maximum contact. If we dump the gristed malt into hot water of 152 degrees F. (or less) temperature the process is accelerated. In five minutes at 148 to 152 degrees we get what is known as an "A" conversion of the starch to malt sugar in solution. In ten minutes at 130 degrees we get another conversion of much of the remaining starch to malt sugar (or maltose) in solution. These temperatures are not absolutes, they are simply the quickest way to get the job done.

The A and B conversions normally convert 70 to 80% of the starch to sugar. When yeast is introduced, the sugar is removed to alcohol and carbon dioxide. The malt enzyme still present then resumes conversion of the rest of the starch. This "secondary conversion" determines your total alcohol yeild.

Yeast

Refer to the flow chart. After the barley starch is converted to sugar, some of it is set aside for a yeast culture and allowed to cool. You can pump in cold water if you are in a hurry. At 90 degrees F., yeast dies but don't put it in until your slurry is below 70 degrees. In a room between 63 and 86 degrees the yeast will raise the temperature of the slurry to 82 or 84 degrees. You don't want it committing suicide. Do not use too much yeast. A few spores between the fingers will do no matter how big a batch you can stick in your basement.

Yeast is basically a fungus (a micro-organism) that eats simple sugars and excretes alcohol and carbon dioxide. You can tell if your yeast is acting on your sugars by carbon dioxide bubbles rising to the top of your slurry. If no bubbles your yeast is not working, it is not producing alcohol, and what you have is just mush. The fermentation process usually takes about 24 hours to take hold so don't panic and throw the yeast mash out fifteen minutes after you inoculate your culture.

Once the bubbles start to rise skim off a cupful and put it in the refrigerator. Be careful not to freeze it. They yeast goes dormant at 62 degrees and can be reactivated above that temperature simply by pouring it in a fresh batch of malt sugar. It is almost the same procedure a housewife uses to make sourdough bread.

The quickest way to get started is to drive down to your local grocery store and buy a small packet of bakers' active dry yeast, known to the scientific crowd as saccharomyces cerevisiae, sometimes known to the housewife under the brand name of "Fleischmann's."

This particular type of yeast can produce 18 to 20% ethanol (sugar alcohol) by volume. At a 20% ethanol concentration yeast dies, which is why alcohol has to be distilled. If anyone ever comes up with a yeast strain that can survive and prosper in an 80% concentration, life would be a whole lot simpler.

Bakers' active dry yeast, when placed in malt sugar of the proper temperature, doubles every two hours through a process known as multi-polar budding. Buds form on the body of the yeast, break off, and form buds of their own, those buds break off, and so on.

This is one reason you never use more than a pinch. Also, put in too much dry yeast and it will raise the temperature of your slurry so quick it will overdose on heat and kill itself.

Yeast growth can also be retarded or prevented by chemicals, heavy metals, and high magnetic fields.

Yeast can grow either in the absence of air (fermentation) or in its presence (respiration, oxidative metabolism). Respiration will give you a higher yield of cells, which means you stir the yeast culture every now and then. Alcohol yield is greater without stirring. Just remember that what you are trying to grow in the yeast culture is yeast and alcohol in your fermenter. The procedures are not the same.

95% of your sugar should go to ethanol and carbon dioxide. You don't want it to go that far in a yeast culture but neither do you want a 70% conversion because of a nitrogen limitation or shortage during fermentation.

Inorganic sources of nitrogen can be utilized for yeast growth.

Ammonia, its sulfate, phosphate salts, and urea can be used in concentrations of up to .06%.

Yeast is sensitive to antibiotics. So don't experiment.

Active dry yeast didn't appear until the 1920's. Up until then yeast cultures were kept in spring houses at low enough temperatures to keep them from growing in a vessel known as a dona.

Yeast occurs in nature. It exists on the surface of fruit and often when the fruit is crushed, cut, or the yeast gains entrance along the stem, fermentation of fruit sugar begins. If you ever see a bird eat a piece of fruit in late fall and then fly off erratically you know what happened: he was just a little tipsy.

Fruit sugar and yeast make wine. Distill it and you have brandy.

Your yeast mash (culture, starter, or whatever you want to call it) should be roughly 2 1/2 lbs of water to 1 lb of malt. You can extend the use of your malt by using 30% barley malt and 70% corn meal.

When the mash is inoculated with your yeast culture you use 2 to 3% yeast starter (the slurry) by volume. That is, throw in a gallon of yeast slurry to every fifty gallons of mash.

Mashing

The process of converting the main body of grain with the aid of the barley malt from starch to sugar is called mashing. Consult the flow chart and follow the directions.

Let's assume you are starting with a bushel or several of corn. At current prices it is one of the most feasible crops of the starches for economic conversion.

The corn should first be sifted through a screen to remove sticks, metal, and trash. A magnet should then be passed over the screen or screens to remove any tramp iron present.

Next run your corn through a coarse grinder or a hammer-mill if you operate on that large a scale. The corn must be cracked or the starch will not be accessible to the malt for conversion to sugar and the kernels will merely swell up by absorbing water. Grind it a little coarser than bread flour.

Feed the corn meal or flour directly into your cooker.

As soon as the hull or protective coating of the kernel is ruptured, it becomes highly susceptible to bacterial action.

Feed 0.5% barley malt into your cooker either at the same time you dump in your corn or mix it in slightly before. The barley malt must be gristed.

A few words of caution at this point.

It is undesirable to store corn meal or flour in large quantities or for any length of time. Grinding corn or any other grain imparts a certain amount of heat to the particles. Heat activates some of the life processes in the particles. If the meal is not used at once there is a heat build-up and the meal becomes caked by the combined action of moisture and heat.

Before the corn is fed into the cooker many of the starch granules are still separated by protective cellulose walls. Heating with water breaks down those walls. The starch then absorbs water, gelatinizes, and goes into solution. Sugar placed in water does almost the same thing by going into solution.

The 0.5% barley mentioned earlier is called pre-malt and simply serves to keep corn from sticking to the bottom of your cooker. It should be added above 152 degrees F. since liquefaction prevails over conversion at higher temperatures.

The corn should be dropped into your cooker as soon as your water hits the boiling point, 212 degrees F., and cooked for not less than thirty (30) minutes.

It is then cooled to 152 degrees F. or slightly less. You can either pump in cool water or allow it to cool naturally. Use a candy thermometer available at any hardware or department store to measure your temperatures.

You should have 25 gallons of water in your cooker for every bushel of corn. Wheat is 17 gallons to the bushel.

Between 148 and 152 degrees F., 6% liquid measure of malted and gristed barley is added to the corn cook. Consult the flow chart. For wheat add only 2% malt and heat at 145 degrees F. for thirty (30) minutes (this is cooking, not conversion time).

The process works as follows:

There are enough enzymes in six pounds of properly prepared barley malt to convert both the barley malt itself and 100 lbs of corn meal to sugar at the proper temperature.

At 148 to 152 degrees an "A" conversion takes place.

At 80 to 130 degrees a "B" conversion takes place.

Any temperature above 152 degrees will kill the enzymes. So be careful. Killing enzymes in pre-malt is alright just in case you were wondering. Just don't kill any you need for your mashing process.

Another word of caution.

The pH of your water is critical for conversion. 7.0 is neutral or distilled water. It is neither acid nor alkaline. Water with a pH factor of above 7.0 is alkaline, below is acid. The desirable pH is between 5.0 and 6.0 for conversion.

Yeast ferments best at 4.0.

Here in Kentucky the water is naturally in that range from our acid soil. In Nebraska it is between 8 and 9, a distillers' nightmare.

pH can be corrected with sulfuric or phosphoric acid (the latter of which is an excellent yeast food). Get some litmus paper and the little color chart that comes with it from a drug store or chemical supply house. The color of litmus paper after it has been dunked in the water should match one of the colors on the chart and tell you what your pH is. If the pH is not high enough add a little sulfuric acid, stir, and test again. The water can be used over and over with fresh batches or corn so it's not as if the sulfuric acid becomes a constant expense.

To test for conversion to sugar you use a sacchrometer or sugar hydrometer, available from any wine shop. It tells you how much sugar you have in solution. Be sure to strain the slurry you are measuring first through a cloth or you will get a reading goofed up by suspended solids.

You should wind up with about 8 to 12 ounces of sugar per gallon of water. 25 gallons of water is 200 lbs, a bushel of corn is 56 lbs, and a bushel of corn will furnish slightly less than 18 lbs of sugar. .12 on the scale (12 ounces) is excellent and .8 (8 ounces) is good for a beginner.

Just because a pro can squeeze 2 1/2 gallons of pure alcohol, 18 lbs of cattle food (distillers dried grains), and the balance in carbon dioxide out of a bushel of corn doesn't mean that you can do exactly the same thing the first time around.

This is only an alcohol motor fuel cookbook, not a course in being a master chef.

Fermenting

This is the great bug bear of the distiller. This is where we separate the folks who are serious from those who are eventually going to have to walk.

This is where we convert sugar to alcohol.

Dump a 2% yeast mash into cooked corn cooled to 70 degrees F. or slightly less.

To recap previous chapters, active fermentation will raise the temperature of the slurry to 82 or 84 degrees. Watch for carbon dioxide bubble activity to step up. At 90 degrees yeast dies, at 62 degrees it goes dormant, and in a 20% concentration of alcohol (ethanol) yeast dies.

The fermentation should take 3 to 4 days.

Test your percentage of alcohol with a hydrometer.

Keep the room temperature between 63 and 86 degrees.

Watch for bubbles of carbon dioxide gas as a ring of froth meets in the center, the whole surface of the mash should become covered with a white creamy foam, and a low hissing sound will begin.

As soon as a crust is formed skim it off and pump the liquid into your still. Leave the settled material in the bottom unless you are pumping it into a column still with stripper plates.

If you allow the crust to fall to the bottom and do nothing, you are going to have another problem.

You will hear the hissing noise again.

The temperature will rise again.

A slight inward movement will be observable.

Floating particles will appear on the surface formed partly into a jelly cake and will become thicker by degrees.

The solution will become nearly transparent.

What you now have cannot be distilled into motor fuel.

You just converted your alcohol into vinegar.

Distilling

Distilling, like making your mash before you put it in your still, requires a certain amount of skill and ability to pay attention to detail. There are certain things to keep in mind.

Don't try to build a complete distillery the first time around, and don't start off with the hardest starch or cellulose to distill that you can find.

Build a simple moonshine still out of a couple of 5 gallon buckets and some copper tubing first. A 5 gallon problem is the same as a 250 gallon one — it is just 50 times easier to deal with.

Try it and see if your still will distill water first. If your cooker isn't sealed properly it won't. Don't waste good mash on a still that is simply going to vent alcohol into the atmosphere before it even gets to your worm or condensor.

The next test is a little expensive but it might save you some grief further on. Once you have satisfied yourself that your still doesn't leak and will distill water, pour a case of beer in it. Beer, incidentally, is what fermented mash is.

Distill the beer and check for proof as it comes out. If you checked your beer for proof before you cooked it and then checked it after you distilled it and did not at least double or triple the proof on a simple moonshine still you had better go back and find out what you did wrong.

If it works with store-bought beer, go to the grocery store and ask for all the rotten fruit they have. Most store owners are only too glad to oblige you. Throw some yeast (at the proper temperature, of course) in the fruit, get it to ferment, and then distill it. Check for proof. Make sure you know what you are doing at this stage.

Perfect the distillation of fruit and then you are ready to go to the malting, mashing, fermenting, and distilling of starch.

Once you become an expert at starch, then you might want to try distilling ethanol from cellulose, an industrial process using acids, or distilling methanol (wood alcohol) from coal.

How to Build and Operate a

Moonshine Still

The still illustrated is about the simplest one you can build that will actually work. About all you will get out of it is enough to run a power lawnmower or other small engines. However, it is good practice.

Let's start with the equipment you will need to build your still.

Two five gallon buckets.

The lid for one of the buckets.

About 21 feet of 3/8 inch copper tubing.

A few feet of plastic hose that will fit over the 3/8 inch tubing.

A garden hose.

A can or jar to catch the alcohol.

Two or three pounds of bread dough.

A source of heat for your cooker. The kitchen stove will do.

The names of the pieces of equipment are given by number matching up with the drawing.

1. Heat source
2. Cooker
3. Cooker lid

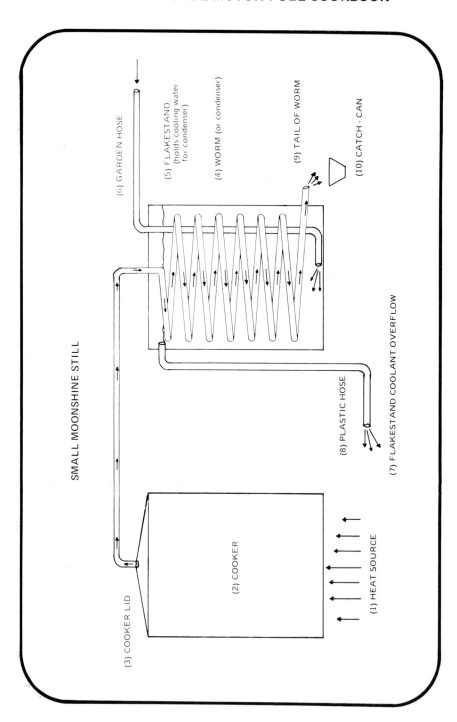

SMALL MOONSHINE STILL

(6) GARDEN HOSE

(5) FLAKESTAND (holds cooling water for condenser)

(4) WORM (or condenser)

(9) TAIL OF WORM

(10) CATCH - CAN

(7) FLAKESTAND COOLANT OVERFLOW

(8) PLASTIC HOSE

(3) COOKER LID

(2) COOKER

(1) HEAT SOURCE

4. Worm (or condensor)
5. Flakestand (holds cooling water for condensor)
6. Garden hose
7. Flakestand coolent overflow
8. Plastic hose
9. Tail of worm
10. Catch-can

The first order of business is to wind the worm, No. 4. Find a round object slightly smaller than the inside diameter of the bucket you are using for a flakestand and simply wrap the copper tubing around it in concentric coils. With larger diameter tubing, up to 1 1/4 inches, you will have to plug one end and pour sand in it to keep if from crimping. The wall strength of the smaller tubing is usually sufficient to prevent crimping.

Once the worm is wound, cut a hole in the bucket for the tail of it, No. 9, and insert it through. Braze or solder the hole shut. Fill it with water to make sure it doesn't leak or you are going to get a whole lot more water than alcohol. Drill another hole at No. 7 for your coolant overflow slightly below where the water level in your flakestand (bucket) is going to be. The plastic pipe or tube, No. 8, is simply to drain the water out from under foot. The garden hose, No. 6, is inserted to the bottom of the flakestand.

Once everything is set up and soldered in place, blow through one end of the worm, 4, and make sure you get some air pressure or a breeze at the other end. If you don't feel any air at the other end, your worm is plugged and even if you do everything else according to directions what you have is no longer a still. It is an explosion waiting to happen!

Assuming you are going to test it to see that it will distill water first, the following directions are for moonshine.

Fill the cooker 2/3rds of the way full with mash. Make a paste of bread dough and water and use it to seal the lid on the cooker and place it around the section of the worm stuck in the lid to keep vapor from escaping. Do a thorough job or

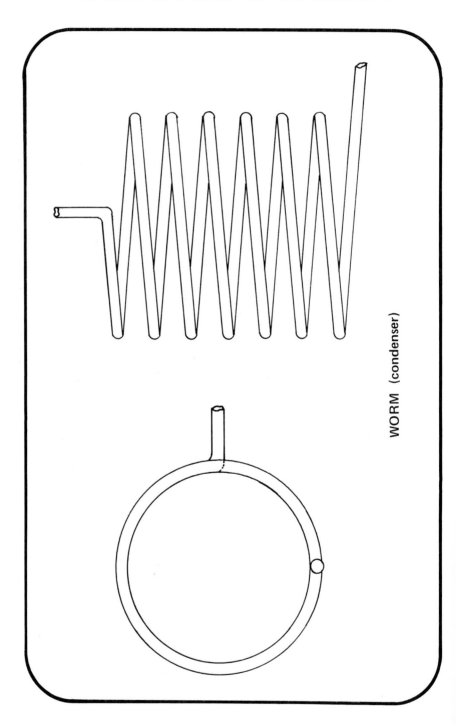

WORM (condenser)

your alcohol and water vapor will escape into the atmosphere. The heat from the cooker causes the bread dough paste (make it thick, like plaster of paris) to harden until it feels like rubber. Most folks don't pay enough attention at this point and their cooker develops vapor leaks, which is why the first run should be made to distill water.

Never — under any circumstances — fill your cooker more than 4/5ths full and not more than 2/3rds if you want to be really safe. A full cooker will "puke" (the liquid passes over into a worm intended only for vapor), the worm becomes foul, or plugged, and the resulting pressure blows the bottom out of your still. Even at 2/3rds full too much heat can cause the same thing.

Fill the flakestand with water. The water should be kept at 60 degrees F. or close to it since if you allow the water to get hotter than the boiling point of alcohol all you will get is distilled water no matter how good your mash was. The alcohol comes out of the tail of the worm as a vapor and escapes into the atmosphere.

Light a fire or turn on the stove under your cooker. If you use a small electric hot plate under the cooker set the cooker dead on top of it — metal to metal — or your heat will get blown away.

Forget to put water in your flakestand and what you have is no longer a still: it is a steam boiler.

Once heat is applied to the still the heat rises to the boiling point of alcohol before it does that of water. Your "first shots" or heads coming out of the tail of your worm will be a stronger proof alcohol than that which follows and the rest of the proof liquor coming out will continue to drop steadily. That is, you may get anywhere from a cup to a pint of 160 to 180 proof alcohol, the next cup or pint will amount to 120 proof, and so on.

You might not even do that.

If so, don't get discouraged.

Simply re-distill it. It is called "doubling" and is done by simply taking everything that came out of the tail of the worm into the catch-can and dumping it back in the cooker, adding enough fresh mash to bring it up to 2/3rds full, and distilling it again. When you have enough skill and practice at doubling, you should be able to raise the distilled matter about 40 proof with each subsequent re-distillation. If you get up to 100 proof in the cooker, it should come out at 140 from the tail of your worm.

There is a point of diminishing returns beyond which it is simply not practical to go. If you get up to 160 proof be satisfied. 200 proof off a moonshine still is almost impossible.

CERTAIN SAFETY PRECAUTIONS SHOULD BE FOLLOWED AND CERTAIN OBSERVATIONS MADE.

Once the still starts working there are several ways to tell if it is behaving properly before the small puff of alcohol vapor from the tail of the worm announces the following arrival of alcohol.

Wrap a rag around your hand and feel the section of the worm between the cooker and the flakestand. The rag is so you won't burn yourself. You should be able to feel the steam traveling.

Sound the worm with an iron-wire rod. If it gives a hollow sound, you are in good shape. If it gives a dull thud, jerk your fire immediately. Your worm is fouled and you are headed for an explosion.

Once the alcohol is running out of the worm tail, your source of heat should be cut back. It takes far less heat to keep your still working than it did to get it working initially. While the still is working, continue listening for a warning sound familiar to all "professional" moonshiners, that is a "BLUP - BLUP - BLUP —" sound from the worm tail. This indicates that too much heat is causing your cooker to work too violently, "puking" liquid, mash, etc. into the worm with the strong possibility of clogging it. With the worm clogged, the fast building pressure has to find another way out — possibily by blowing up in your face. Upon hearing the "BLUP - BLUP - BLUP —" quickly turn down or temporarily remove the heat until the distilling process has stabilized.

Always remember that the still is just as dangerous as a steam boiler it if goes into orbit.

Watch for water vapor rising off the flakestand. Another sign the still is running true.

Stick a thermometer in the flakestand water every now and then to check the temperature. When the cooling water starts getting too warm all you have to do is turn on the garden hose. The cooling water from the garden hose will force the hot water — which will be on the top, both because heat rises and the worm is the hottest when it first enters the water — out the exhaust.

Testing for Proof

Alcohol won't tell you what proof it is just by looking at it. When the stuff comes out of the still it looks just like water. However, since alcohol weighs 6.6 lbs to the gallon as opposed to 8 lbs per gallon for water there is a way to check.

Specific gravity.

A float with a weight on one end will sink further in alcohol than it will in water. Put a series of measurements on the float and you have a hydrometer. It will measure "0" when it is floating in water and "200" when it is floating in pure alcohol. The scale starts at "0" and goes up towards the ceiling.

Any wine shop can order one for you.

Temperatures of all wine shop hydrometers are calibrated for liquid of 60 degrees F. temperature. If you want to be really picky you can order the Gauging Manual from the Internal Revenue Service that has columns telling you exactly what the proof is by matching up hydrometer readings with temperature charts.

It allows you to tell proof to a tenth of a per cent. Assuming, however, that you know enough to use a thermometer to take the temperature of the liquid without putting your hand on the test tube and heating up the reading.

A Larger Moonshine Still

If you have ever traveled through the backroads of Darkest Nebraska you may have noticed the tremendous number of empty drums and buckets strewn all over the local farms, businesses, and so on.

Hardware stores even carry fifty foot coils of copper tubing in plain view.

A moonshiner's paradise.

Once you get past the five gallon stage and have a fairly good idea of what you are doing, it is time to move on to bigger and better things. A 55 gallon still is the same as a 5 gallon bucket still so we will pass that and get right into serious business.

Like a 250 gallon cooker. Keep a sharp eye out and you can probably pick one of these up at the city dump.

Follow the illustration. A lot of things are indentical to the 5 gallon cooker. Some aren't. Everything is listed and numbered regardless to avoid confusion.

1. Heat source. Obviously this monstrosity it not going to fit on your kitchen stove.

2. Cooker

3. Screw cap

4. Drain faucet

5. T passage to two worms, two flakestands

6. Cooling water funnel and pipe welded to inside of 55 gallon flakestand. Stick your hose in here, carry buckets of water, or however you want to do it.

7. Flakestand

8. Worm. One on each side.

9. Tail of worm

10. Exhaust pipe for flakestand

The purpose of this arrangement is speed and volume production. Since copper tubing of over 1 1/4 inch diameter will not cool the vaporized alcohol sufficiently more than one worm is run off the still. Production is doubled. The T tube is never less than two (2) inches and always the same diameter going in both directions, otherwise one worm will rob the other.

The faucet, 4, is to drain the cooker after all the alcohol has been distilled out and to make room for a fresh batch of mash. The screw cap, 3, is to avoid all the mess with the bread dough and allow fresh mash to be funneled into the cooker.

Out of this size cooker you should be able to get 5 to 10 gallons of motor fuel on the first run of 160 proof. Use your hydrometer. Double the remainder just as you would on the five gallon bucket.

For all stills use the formulas for corn and water given elsewhere in this book. That is, 25 gallons of water to a bushel of corn. Since you cannot fill any still to the top figure 200 gallons of water in this one to 8 bushels of corn, which should eventually result in at least 16 gallons of motor fuel.

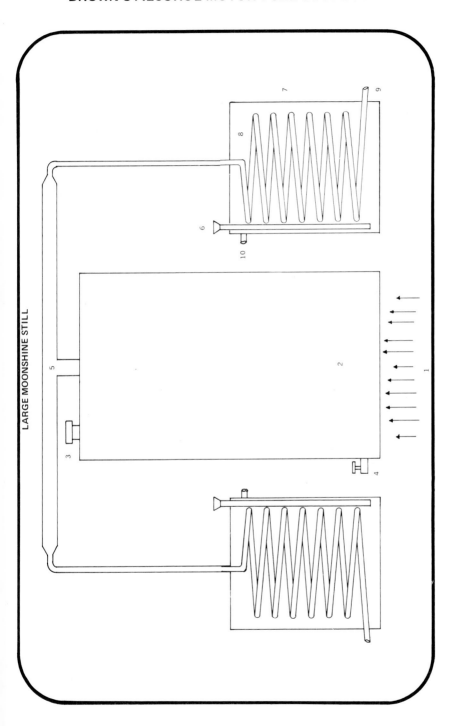

LARGE MOONSHINE STILL

The Doubler

The reason this attachment between the still and the flakestand is so named is that it often doubles the proof of the alcohol without you having to pour it back in the still.

It works like this:

Alcohol and water vapor travel up the pipe, No. 1.

This same vapor passes through at least five inches of water in the doubler, 2. When the water rises to the boiling point of alcohol, the alcohol vapors pass into the worm, No. 4, and the water above the level of the return pipe's mouth, No. 3, returns to the bottom of the still.

All industrial stills in use today are based on this principle. A series of doublers stacked on top of each other, called a column.

The first still capable of producing industrial alcohol in this country in 1801 was called Adam's Still and was merely two doublers hooked up in series. He did have a few problems with temperature control of the water in the doublers, high vapor pressure of the boiling liquid, and a few other items you can solve with a trip or two to the hardware store for pressure and temperature gauges, pumps, and a few other items not available in 1801.

Though you would probably be better off building a column.

Some large distilleries use two condensers (worms) on each side of a doubler to control the exact proof. Heat in the doubler is kept constant by control of the air temperature in the doubler as opposed to letting the alcohol vapors pass through the water. The accumulated water is not automatically drained back into the still and may rise from six inches in the morning to six feet at night.

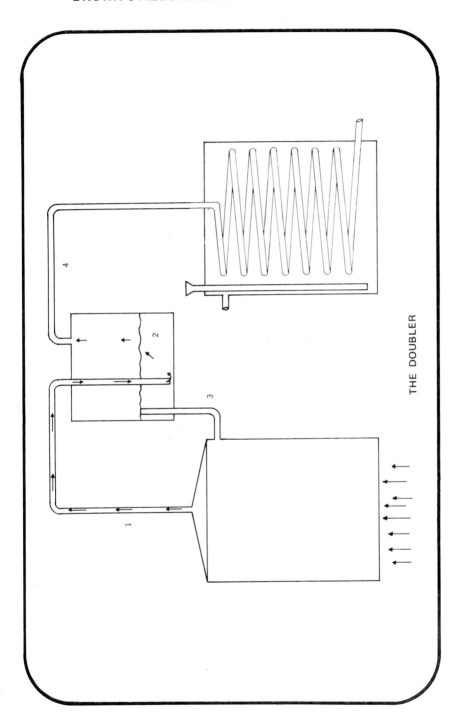

THE DOUBLER

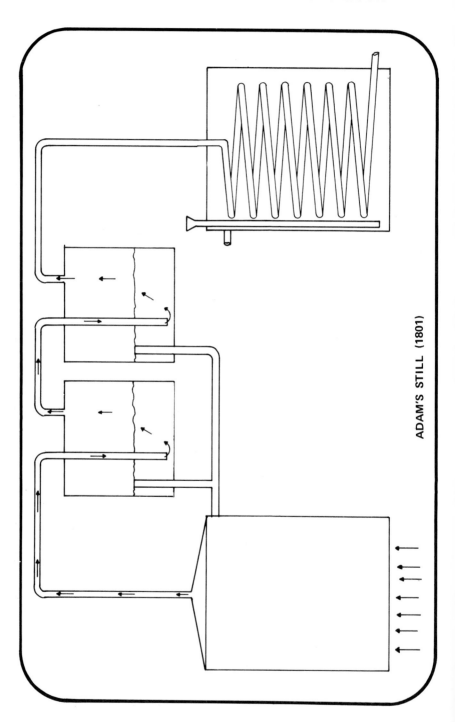

ADAM'S STILL (1801)

The Column

The column is merely a series of doublers placed on top of each other. The easiest way to make a column is to use plumbing supplies and then either weld or bolt everything together.

The way it works is as follows:

Fresh mash from the pipe 1 is fed into the still 2. Before the still is heated and the alcohol and water vapors allowed to rise through the pipe 3, the still is "charged" with fresh water by opening the valves in 4 and inserting a hose at the top. Water is allowed to achieve the height of the down pipes 5 and then runs down to the next level. The down pipe extends seven inches up and sixteen inches down into the next level, two inches from the floor of the chamber. This down pipe is sometimes also called a "plunger."

The center pipe 6 is four inches in diameter and eight inches high. The cap 7 that goes over the center pipe and is welded to the floor of the chamber is twelve inches in diameter at the top and sixteen inches at the bottom. The bottom of the cap is two inches from the floorplate.

The charge hole 8 with screw cap is for cleaning the still out. A safety valve like those used on a steam boiler should be provided somehwere on the still to prevent explosion.

The column section nine is made out of flanged pipe and bolted together. With the price of flanged pipe these days, you might be better off cutting and welding and drilling your own flanges. Use heat and alcohol resistant gel as a sealer.

The floor plate 10 is cut out of suitable flat stock and holes drilled to accomodate the holes in the flanged pipe.

The flanged pipe is eighteen inches in length and three feet in diameter. The down pipes or plungers are two inches in diameter.

Up to a dozen sections can be bolted on top of each other to make a column eighteen feet high capable of producing over 3 gallons a minute of 160 or better proof alcohol. The measurements are all proportional and can be scaled up or down.

But don't make your still too skinny.

A twelve inch diameter still will produce about twenty gallons of alcohol an hour. One three feet in diameter will produce about two hundred gallons an hour. The larger a still is, the more effective it is. A small column will chew up half as many BTU's as it produces in alcohol. A large industrial still will use only one steam horsepower hour to produce twenty gallons of alcohol.

When the proof of the finished product starts dropping too low, drain the cooker, pump in fresh mash (the pressure in the feed pipe 1 must be greater than in the cooker 2. Set the understrength alcohol in a separate tank and re-distill it. Understrength alcohol is called "low wines."

Do not dump in the dregs from your mash unless you use stripper plates between the still and the column unless you want to clog it up and have a blow-up.

This column is also referred to as a rectifying column.

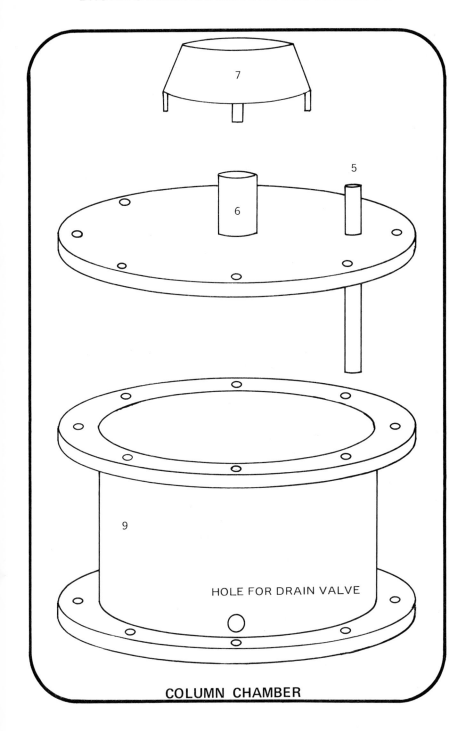

COLUMN CHAMBER

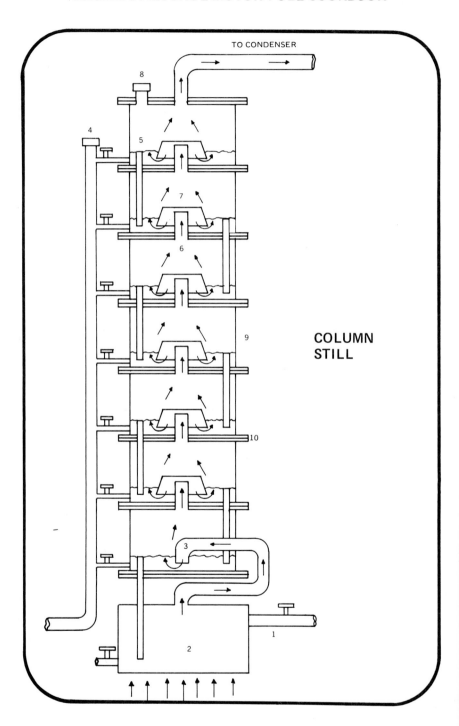

TO CONDENSER

COLUMN
STILL

Stripper Plates and Beer Stills

One of the problems with a conventional column or still is that the dregs or solid matter left over after fermenting have nowhere to go. They just squat in the bottom of the mash tubs or fermentors and look at you. Clean-up is a mess.

This is solved by the use of stripper plates.

Slurry is fed into the beer still via a feed pipe 1. This slush travels across a perforated plate 2. Steam pressure of 3 to 4 lbs per square inch in the still keeps the slurry from falling through the perforations in the plate.

When the slurry reaches the level of the downpipe 3 it falls down to the next level or stripper plate. When it hits the liquid seal or cup 4, it forms a barrier of liquid to keep vapor from traveling up the downpipe before it spills over onto the next plate.

Since the still is heated by steam passing through or an external source of heat the alcohol and water vapor rises, passes through the holes in the stripper plates, condenses partially, releases heat, and vaporizes an equivalent of liquid on that plate.

The process is repeated on several plates until all the alcohol is "stripped" from the plates.

Vaporized alcohol and water then enter the rectifying

column. Water and suspended solids then leave the base of the beer still and are either pumped or gravity fed to an evaporator.

The water is then removed and the result is known as distillers dried grains, a very high quality of cattle food.

Beer is fed three of four plates down to avoid an accident clogging the rectifying column. A reflux line from the condenser keeps the top three of four stripper plates wet. The slurry being pumped to the evaporator is checked peridically for alcohol content with a hydrometer. If a positive reading results, you are either going to have to add more stripper plates or pay closer attention to your temperature control.

The distance between stripper plates is the same as it is for the rectifying column. The liquid seal is four inches high and the down pipe three inches above the surface.

The liquid reflux line from the condenser is bent like a plumbers' elbow to act as a liquid seal from the condenser to the beer still.

The vapor line to the condenser is wrapped around the incoming beer feed (or vice versa) in a device called a preheater to warm the incoming mash and help cool the outgoing vapor.

If you have followed directions up to this point you are (or should be) congratulated.

You have just built a distillery.

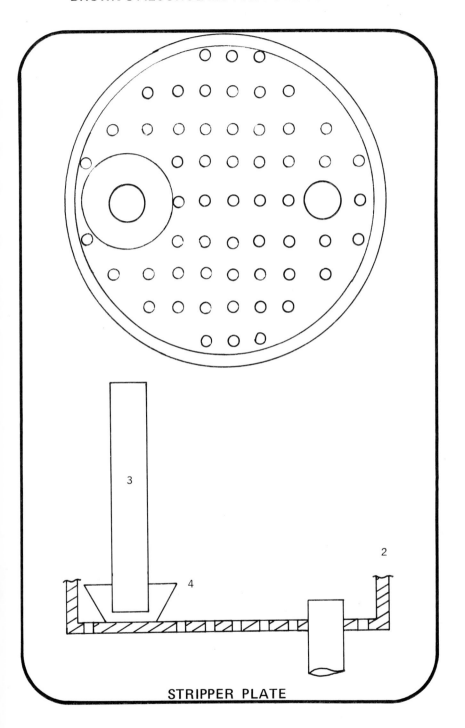

STRIPPER PLATE

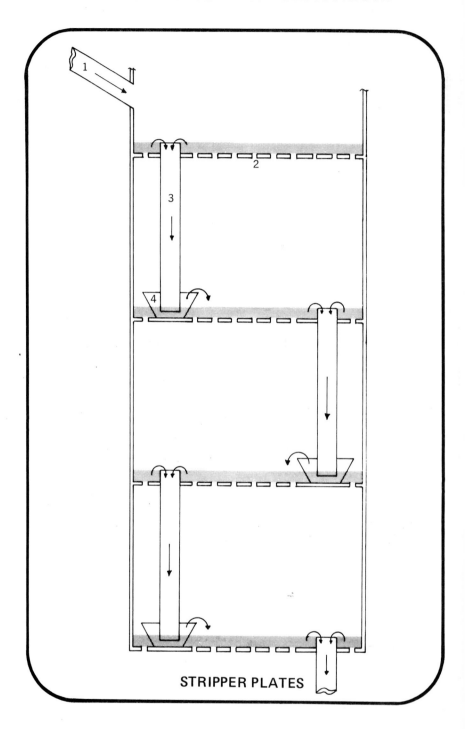

STRIPPER PLATES

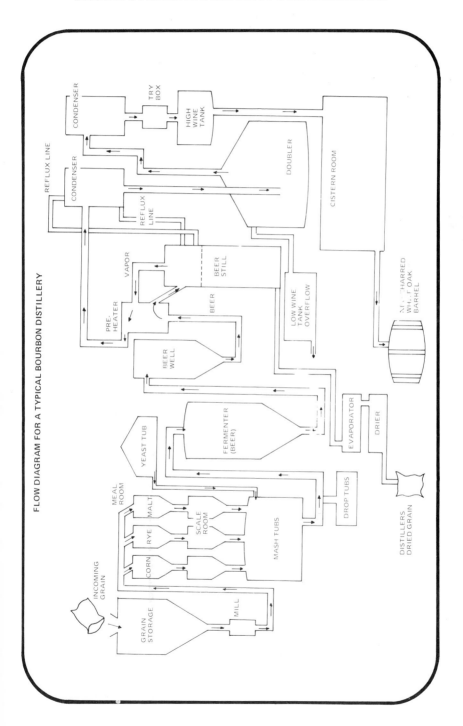

FLOW DIAGRAM FOR A TYPICAL BOURBON DISTILLERY

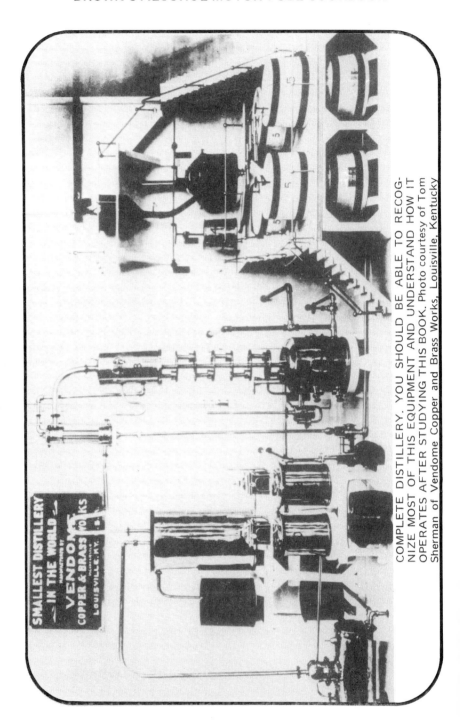

COMPLETE DISTILLERY. YOU SHOULD BE ABLE TO RECOGNIZE MOST OF THIS EQUIPMENT AND UNDERSTAND HOW IT OPERATES AFTER STUDYING THIS BOOK. Photo courtesy of Tom Sherman of Vendome Copper and Brass Works, Louisville, Kentucky

J. D. COOK OF AUSTIN - NICHOLLS DISTILLERY NEXT TO GRAIN STORAGE FUNNELS. GRAIN IS STORED UNDERNEATH HIS FEET.

GRAIN IS GROUND INTO FLOUR VIA HAMMERMILL. A COARSE
ELECTRIC GRINDER WILL WORK IN YOUR KITCHEN.

GROUND GRAIN IS DUMPED INTO MASH TUB(s). A LID IS NOT ESSENTIAL FOR KITCHEN COOKING.

FERMENTING TUB FOR MASH.

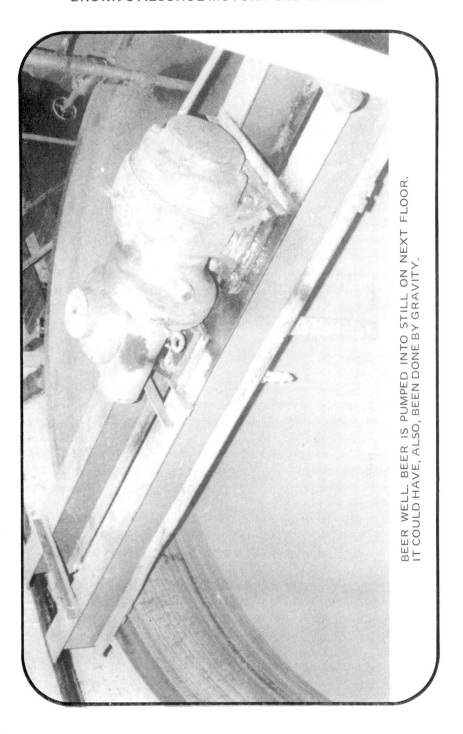

BEER WELL. BEER IS PUMPED INTO STILL ON NEXT FLOOR.
IT COULD HAVE, ALSO, BEEN DONE BY GRAVITY.

STRIPPER - PLATE SECTION OF COLUMN BEER STILL. PIPE WITH HAND ON IT IS FOR FLUSHING (cleaning) STILL. DOORS ARE FOR CLEANING STRIPPER PLATES.

RECTIFYING SECTION OF SAME STILL. LARGE PIPE GOES TO CONDENSER. SMALL PIPE IS REFLUX LINE.

BEER FEED PIPE TO FOURTH STRIPPER PLATE DOWN. STILLS LIKE THIS ONE ARE MADE OF COPPER SIMPLY FOR TASTE — STAINLESS STEEL & OTHER MATERIALS MAKE WHISKEY TASTE LIKE BOILED CABBAGE.

TOPS OF CONDENSERS. THIS SIZE (condensers) USE A SERIES OF STRAIGHT PIPES (similar to a car radiator) INSTEAD OF A COILED PIPE (like a moonshine still).

COMMERCIAL DOUBLER. SMALL PIPE IS INCOMING, LARGE PIPE IS OUTGOING.

DISTILLERS DRIED GRAINS — 26% PROTEIN CATTLE FEED.

TRUCK BEING LOADED WITH DISTILLERS DRIED GRAINS FROM EVAPORATOR ROOM AT AUSTIN - NICHOLLS DISTILLERY, LAWRENCEBURG, KENTUCKY.

Gasohol

Don't try home distilled alcohol in your gas tank mixed with gasoline unless you add another couple of steps to your operation to get anhydrous or 200 proof alcohol.

Any water in the alcohol or the gasoline will cause the blend to separate, which means you will have:

Hard starting

Hesitation on accleration

Stalling at stoplights

And other problems.

At 195 proof alcohol and water distill at the same temperature. This is called azeotropic mixture.

The last 5 proof is a real bear.

For the alcohol to be separated from that last 2 1/2 per cent of water the water must be presented with something it likes better than alcohol.

To get 200 proof alcohol pour 190 proof or better into a bucket of quick lime or caustic soda. Let it soak overnight in a sealed container.

Distill the mixture the next day. The quick lime or caustic soda will hold the water and release the alcohol. Just make sure you put your 200 proof mixture in an airtight jar or the proof will start dropping as it sucks moisture out of the air.

Use just a pint or two of this mixture in your gas at a time until you feel relatively certain it has soaked up all the condensation in your tank and run it through the engine.

A Solar Still

Basically, a solar alcohol still is the same thing as a solar water still. The primary difference is in the liquid you put in it, not the plumbing.

The solar still works just like a car with all the windows rolled up on a hot day.

Glass windows in a car or a solar still permit the short wave length of the sun's ultra-violet rays to pass through. The glass does not permit the long wave length of heat radiation to pass back through.

The heat is then trapped. Sometimes this is known as a "greenhouse effect."

There are a number of disadvantages to using solar stills. It is not my favorite way to do things.

The first major disadvantage is cost. The money required to set up a halfway efficient operation is going to go way beyond that you will have to put up for even a small sized column with stripper plates on a per gallon basis. A whole barn roof covered with solar stills might — if you are lucky and have a lot of hot, still days — yeild 10,000 gallons of alcohol in a four month period.

A twelve-inch diameter column will yeild twenty gallons an hour running an natural gas, two hundred gallons a day, or

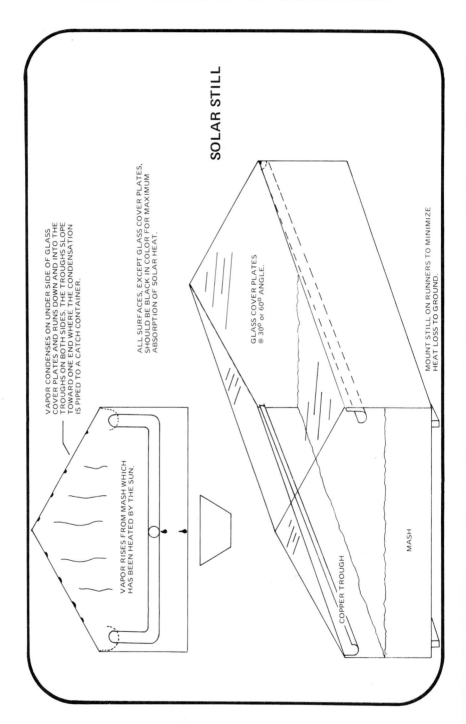

SOLAR STILL

VAPOR CONDENSES ON UNDER SIDE OF GLASS COVER PLATES AND RUNS DOWN AND INTO THE TROUGHS ON BOTH SIDES. THE TROUGHS SLOPE TOWARD ONE END WHERE THE CONDENSATION IS PIPED TO A CATCH CONTAINER.

ALL SURFACES, EXCEPT GLASS COVER PLATES, SHOULD BE BLACK IN COLOR FOR MAXIMUM ABSORPTION OF SOLAR HEAT.

GLASS COVER PLATES @ 30° or 60° ANGLE.

MOUNT STILL ON RUNNERS TO MINIMIZE HEAT LOSS TO GROUND.

VAPOR RISES FROM MASH WHICH HAS BEEN HEATED BY THE SUN.

COPPER TROUGH

MASH

AUTHOR'S SOLAR STILL.

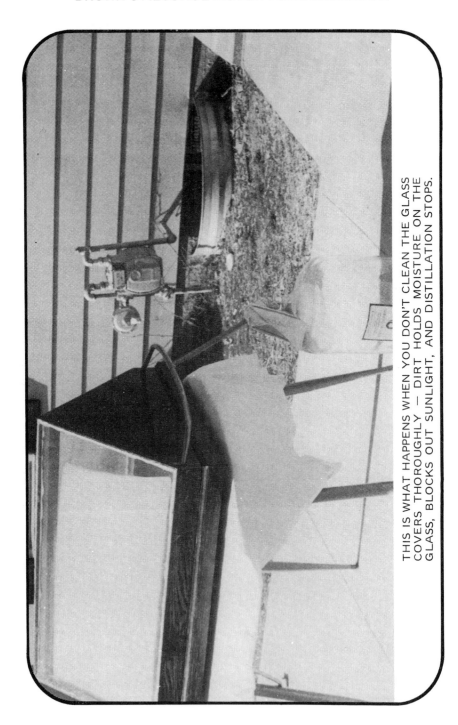

THIS IS WHAT HAPPENS WHEN YOU DON'T CLEAN THE GLASS COVERS THOROUGHLY — DIRT HOLDS MOISTURE ON THE GLASS, BLOCKS OUT SUNLIGHT, AND DISTILLATION STOPS.

the same amount of fuel in one-third the time. Regardless of the weather.

Another drawback is temperature control. You can turn steam, natural gas, or even the wood on a moonshine still up or down. In the case of a small still banking a fire when the charge starts to run is almost a necessity. The sun is pretty much going to do what it wants to. You would wind up getting anything from 100 proof first shots all the way down to distilled water with less than the alcohol content of beer. The only way I know of to minimize this problem is to hook solar stills up in series, put the mash in still number one, run a line from it into the next still filled with caustic soda on hand that has to be dried out.

However, for die hard solar energy fans, you do it this way:

The bottom layer of the still is usually wood. The next layer up is three inches or so of fiberglass to insulate and retain heat. The third layer up is black painted or anodized metal used to absorb the sunlight once the rays get past the glass. The fourth layer up is your fluid which is heated by conduction of the heat from the metal. The fifth layer up is the empty space between the fluid and the glass where the vapors rise and condense against the glass which in turn slides down into the runners.

The glass should always be set at a 30 or 60 degree angle since light absorption at those angles is twice what it is at 45 degrees. Keep the glass clean or you will wind up with moisture sticking to the dirt and shutting out the sunlight.

There are a number of interesting projects that could be undertaken on solar stills that are beyond the scope of this book though I will mention them in case you care to experiment in such directions.

Liquid vaporizes more readily in a vacuum. A vacuum pump on a solar still might decrease the amount of heat required to vaporize the alcohol and water vapor.

The collection of heat via the fresnel lens.

The utilization of a solar furnace, fresnel lens, or parabolic collectors to heat the water in a flash boiler. A flash boiler would give the steam necessary for corn cookers. Other-

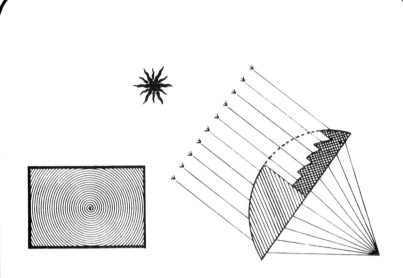

FRESNEL LENS

A FRESNEL LENS, AS OPPOSED TO A CONVENTIONAL LENS, IS VERY THIN, MADE FROM FLEXIBLE PLASTIC MATERIAL AND IS USUALLY SQUARE OR RECTANGULAR IN SHAPE. THESE FEATURES ALLOW IT TO BE INSTALLED IN APPLICATIONS WHERE A CONVENTIONAL LENS WOULD BE IMPRACTICAL. A LARGE SQUARE PIECE CAN BE INSTALLED AND, SINCE THE RADIAL GROVES ALL HAVE THE SAME FOCAL POINT, IT WILL GATHER MORE SOLAR ENERGY THAN A ROUND CONVENTIONAL LENS THAT WOULD FIT THE SAME AVAILABLE AREA. ONE SOURCE OF LARGE FRESNEL LENS KNOWN TO THE AUTHOR IS: Creative Conservation Service, 4375 Yale Station, New Haven, CT 06520.

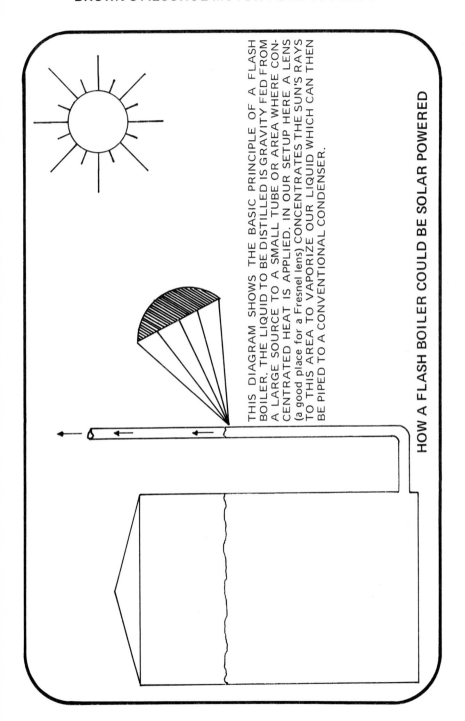

THIS DIAGRAM SHOWS THE BASIC PRINCIPLE OF A FLASH BOILER. THE LIQUID TO BE DISTILLED IS GRAVITY FED FROM A LARGE SOURCE TO A SMALL TUBE OR AREA WHERE CONCENTRATED HEAT IS APPLIED. IN OUR SETUP HERE A LENS (a good place for a Fresnel lens) CONCENTRATES THE SUN'S RAYS TO THIS AREA TO VAPORIZE OUR LIQUID WHICH CAN THEN BE PIPED TO A CONVENTIONAL CONDENSER.

HOW A FLASH BOILER COULD BE SOLAR POWERED

wise you have to sprout and grind every grain of corn you put in your solar still. Corn meal cooked at 360 degrees has to be cooked for only one minute. A large fresnel lens can generate temperatures of up to 2500 degrees F.

On the assumption that a picture is worth a thousand words, take your pick. The only comprehensive work on flash boilers I know of is sold by Caldwell Industries in Luling, Texas.

Dealing With Big Brother

The hardest part of any honest endeavor these days seems to be keeping bureaucrats out of the way. Repeal all the restrictions on alcohol production and this country could be energy self-sufficient within a matter of weeks.

Don't hold your breath, though. As long as Congress is stuffed with lawyers all they are going to do is pass laws by the truckload. It is all they know how to do.

Vote for a farmer, a machinist, a mechanic, or possibly a moonshiner next time.

As long as lawyers infest the Congress you will just have to live with the paperwork.

To manufacture alcohol commercially, the Internal Revenue Department requires a tax (naturally) stamp on every still plus a large bond just to insure that you won't sell your product or "dranken likker" thereby nerfing the Government parasites out of their $10.50 a gallon tax that costs $10.00 to collect. Alcohol over 185 proof must be denatured by adding one gallon of gasoline to every hundred gallons of alcohol.

At the time of this writing, the BATF (Bureau of Alcohol, Tobacco and Firearms) division of the Treasury Department, is licensing some home stills for experimental use.

Typically, Government controls, licensing, etc. change often and vary from area to area. For legally making alcohol the reader is advised to contact the nearst BATF office for the latest information.

With the gasoline shortage and high prices developing as they are, there is a strong possibility that masses of certain civil disobedient types will begin cooking alcohol motor fuel regardless of what the government tries to do about it.

Objections and Comments

About Farm Alcohol

As a Motor Fuel

I don't have any.

Do you?

Seriously, most of the objections to the use of crops as a motor fuel come from:

Oil companies. Hardly an unbiased source.

Multi-million dollar distillation plants, which would collapse almost immediately following the average farmer building his own still and distilling his own crops.

Economics experts. Half of whom can't even balance their own checkbooks!

I have never heard an American farmer say, "We can't do it."

In all fairness to the members of Congress, a few of them do seem to have their hearts in the right place.

Unfortunately, less than a handful of them know the difference between a venturi tube and an accelerator pump and they keep calling oil company employees with degrees in mechanical engineering to "tell them like it is."

No wonder we have an energy problem.

Appendix

The same basic rules apply for almost all types of fruit. It is already a sugar (fructose) and does not need to be mashed.

The formulas for corn and wheat have already been given.

Potatoes are steamed at 45 to 60 lbs per square inch in a steaming vat and then mashed in the same fashion as corn. The steaming breaks down the starch. Only crushed potato pulp is used. Seven pounds of malt is used to one hundred pounds of potatoes. Cooking is done at 140 degrees for three to four hours.

If you live in Idaho you might want to malt your potatoes and bypass the barley malt.

The processes required for sugar beets would fill up a book to be completely accurate. The juice can be squeezed out hydraulically and a small quantity of sulfuric acid is added. Beets are famous for clogging up your equipment.

Molasses is the most common source of ethanol overseas. A reprint of the chapter on molasses follows from the 1906 work of Charles Wright, DISTILLATION OF ALCOHOL.

ALCOHOL FROM MOLASSES

Another common source of alcohol is molasses. Molasses is the uncrystallizable syrup which constitutes the residium of the manufacture and refining of cane and beet sugar. It is a dense, viscous liquid, varying in color from light yellow to almost black, according to the source from which it is obtained; it tests usually about 40° by Baume's hydrometer. The molasses employed as a source of alcohol must be carefully chosen; the lightest in color is the best, containing most uncrystallized sugar. The manufacture is extensively carried on in France, where the molasses from the beet sugar refineries is chiefly used on account of its low price, that obtained from the cane sugar factories being considerably dearer. The latter is, however, much to be preferred to the former variety as it contains more sugar. Molasses from the beet sugar refineries yields a larger quantity and better quality of spirit than that which comes from the factories. Molasses contains about 50 per cent of saccharine matter, 24 per cent of other organic matter, and about 10 per cent of inorganic salts, chiefly of potash. It is thus a substance rich in matters favorable to fermentation. When the density of molasses has been lowered by dilution with water, fermentation sets in rapidly, more especially if it has been previously rendered acid. As, however, molasses from beet generally exhibits an alkaline reaction, it is found necessary to acidify it after dilution; for this purpose sufluric acid is employed, in the proportion of about 4 1/2 lbs. of the concentrated acid to 22 gallons of molasses, previously diluted with eight or ten volumes of water. Three processes are thus employed in obtaining alcohol from molasses; dilution, acidification, and fermentation. The latter is hastened by the addition of a natural ferment, such as brewer's yeast. It begins in about eight or ten hours, and lasts upwards of 60.

BEET SUGAR MOLASSES

The first step in the process of rendering the molasses fermentable is to mix the molasses with water, to a certain dilution. This may be done by hand, but preferably it is performed in a vat provided with stirring or agitating mechanism, such as will effectually mix the water with the viscid syrup, and whereby also the wash may be thoroughly agitated and aerated.

There are numerous forms of mixing vats, all working however, on the principle shown in Fig. 1. In this, the vat A is provided with a central shaft C carrying radial mixing blades E. This shaft is driven by bevel gears D, F. As the rotation of these blades would merely tend to create a rotary current of molasses and water, and not mix them, some means should be used for impeding and breaking up this current. To that end the cover is provided with downwardly projecting rods I which create counter currents, and thoroughly intermingle the two liquids. Another and even better form of mixer consists of a tank into the lower portion of which enters a perforated pipe of relatively large diameter. This is provided at the end with an air entrance and a steam injector. The injected steam draws in air and the steam and air are forced under pressure into the vat, thus diluting the contained molasses, agitating it and thoroughly aerating it.

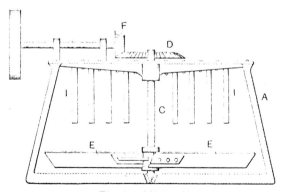

Fig. 1 — Mixing Vat.

The molasses as it comes from the sugar house may contain anywhere from 30 to 45% of sugar, and this should be diluted with water to a concentration of 16% to 18% of sugar.

The density of the wash after "setting up" is 1060. It is to be noted that though with improved apparatus a wash as concentrated at 12 degrees or 15 degrees Baume may be worked; yet where simple apparatus is used six degrees or eight degrees is better and much more favorable to rapid and complete fermentation.

After setting up, one gallon of strong sulphuric acid and 10 lbs. of sulphate of ammonia are added for each 1000 gallons of wash. This neutralizes the alkaline carbonates in the beet juice which would otherwise retard fermentation, and it assists the yeast to invert the cane sugar as formerly described. The addition of ammonia is in order to give food to the yeast and obtain a vigorous fermentation.

The yeast used for fermenting molasses is prepared either from malt or grain and is used as concentrated as possible.

The "pitching" temperature of a molasses wash varies with the concentration of the wash, being higher for strongly concentrated solutions than for weak ones. When the wash test as high as 12 degrees Baume, fermentation begins at about 77° F. and is raised during fermentation to 85° or 90° F. A temperature around 82° is best on the average as this is most conducive to the growth of yeast.

Where the vats are large and the syrup considerably diluted the temperature rises very quickly and must be moderated by passing a current of cold water through a coil of pipe on the bottom of the vat.

FERMENTING RAW SUGAR

This is accomplished by dissolving the sugar in hot water, then diluting it, and then adding a ferment, — fermentation being aided by adding sulphuric acid to the diluted molasses, in the proportion of one-half to one pound of acid to every hundred pounds of pure sugar used.

The wash is pitched with compressed yeast in the proportion of 2 1/2 to 8% of the weight of the sugar used. The pitching temperature is from 77° to 79° F., and the period of fermentation is 48 hours.

CANE SUGAR MOLASSES

Besides the molasses of the French beet sugar refineries, large quantities result from the manufacture of cane sugar in Jamaica and the West Indies. This is entirely employed for the distillation of rum. As the pure spirit of Jamaica is never made from sugar, but always from molasses and skimmings, it is advisable to notice these two products, and together with them, the exhausted commonly called dunder.

The molasses proceeding from the West Indian cane sugar contains crystallizable and uncrystallizable sugar, gluten, or albumen, and other organic matters which have escaped separation during the process of defecation and evaporation, together with saline matters and water. It therefore contains in itself all the elements necessary for fermentation, i.e., sugar, water, and gluten, which latter substance, acting the part of a ferment, speedily establishes the process under certain conditions. Skimmings comprise the matters separated from the cane juice during the processes of defecation and evaporation. The scum of the clarfiers, precipitators, and evaporators, and the precipitates in both clarifiers and precipitators, together with a proportion of cane sugar mixed with the various scums and precipitates, and the "sweet-liquor" resulting from the washing of the boiling-pans, etc., all become mixed together in the skimming-receiver and are fermented under the name of "skimmings." They also contain the elements necessary for fermentation, and accordingly they very rapidly pass into a state of fermentation when left to themselves; but, in consequence of the glutinous matters being in excess of the sugar, this latter is speedily decomposed, and the second, or acetous fermentation, commences very frequently before the first is far advanced. Dunder is the fermented wash after if has undergone distillation, by which it has been deprived of the alcohol contained. To be good, it should be light, clear, and slightly bitter; it should be quite free from acidity, and is always best when fresh. As it is discharged from the still, it runs into receivers placed on a lower level, from which it is pumped up when cool into the upper receivers, where it clarifies, and is then drawn down into the fermenting cisterns as required. Well-clarified dunder will keep for six weeks without injury. Good dunder may be considered to be the liquor, or "wash," as it is termed, deprived by distillation of its alcohol, and much concentrated by the boiling it has been subjected to; whereby the substances it contains, as gluten, gum, oils, etc., have become, from repeated boilings, so concentrated as to render the liquid mass a highly aromatic compound. In this state it contains at least two of the elements necessary for fermentation, so that, on the addition of the third, viz, sugar, that process speedily commences.

The first operation is to clarify the mixture of molasses and skimmings previous to fermenting it. This is performed in a leaden receiver holding about 300 or 400 gallons. When the clarification is complete, the clear liquor is run into the fermenting vat, and there mixed with 100 or 200 gallons of water (hot, if possible), and well stirred. The mixture is then left to ferment. The great object that the distiller has in view in conducting the fermentation is to obtain the largest possible amount of spirit that the sugar employed will yield, and to take care that the loss of evaporation or acetification is reduced to a minimum. In order to ensure this, the following course should be adopted. The room in which the process is carried on must be kept as cool as it is possible in a tropical climate; say, 75° to 80° F.

Supposing that the fermenting vat has a capacity of 1000 gallons, the proportions of the different liquors run in would be 200 gallons of well-clarified skimmings, 50 gallons of molasses, and 100 gallons of clear dunder; they should be well mixed together. Fermentation speedily sets in, and 50 more gallons of molasses are then to be added, together with 200 gallons of water. When fermention is thoroughly established, a further 400 gallons of dunder may be run in, and the whole well stirred up. Any scum thrown up during the process is immediately skimmed off. The temperature of the mass rises gradually until about 4^0 or 5^0 above that of the room itself. Should it rise too high, the next vat must be set up with more dunder and less water; if it keeps very low, and the action is sluggish, less must be used next time. No fermenting principle besides the gluten contained in the wash is required. The process usually occupies eight or ten days, but it may last much longer. Sugar planters are accustomed to expect one gallon of proof rum for every gallon of molasses employed. On the supposition that ordinary molasses contains 65 parts of sugar, 32 parts of water, and three parts of organic matter and salts, and that, by careful fermentation and distillation, 33 parts of absolute alcohol may be obtained, we may then reckon upon 33 lbs. of spirt, or about four gallons, which is a yield of about 5 2/3 gallons of rum, 30 per cent over-proof, from 100 lbs. of such molasses.

The following process is described in Deer's work on "Sugar and Sugar Cane."

"In Mauritius a more complicated process is used; a barrel of about 50 gallons capacity is partly filled with molasses and water of density 1.10 and allowed to spontaneously ferment; sometimes a handful of oats or rice is placed in this preliminary fermentation. When attenuation is nearly complete more molasses is added until the contents of the cask are again of density 1.10 and again allowed to ferment. This process is repeated a third time; the contents of the barrel are then distributed between three or four tanks holding each about 500 gallons of wash of density 1.10 and 12 hours after fermentation has started here, one of these is used to pitch a tank of about 8,000 gallons capacity; a few gallons are left in the pitching tanks which are again filled up with wash of density 1.10 and the process repeated until the attenuations fall off, when a fresh start is made. This process is very similar to what obtains in modern distilleries save that the initial fermentation is adventitious.

"In Java and the East generally, a very different procedure is followed. In the first place a material known as Java, or Chinese, yeast is prepared from native formulae; in Java, pieces of sugar cane are crushed along with certain aromatic herbs, amongst which galanga and garlic are always present, and the resulting extract made into a paste with rice meal; the paste is formed into strips, allowed to dry in the sun and then macerated with water and lemon juice; the pulpy mass obtained after standing for three days is separated from the water and made into small balls, rolled in rice straw and allowed to dry; these balls are known as Raggi or Java yeast. In the next step rice is boiled and spread out in a layer on plantain leaves and sprinkled over with Raggi, then packed in earthware pots and left to stand for two days, at the end of which period the rice is converted into a semi-liquid mass; this material is termed Tapej and is used to excite fermentation in molasses wash. The wash is set up at a density of 25° Balling and afterwards the process is as usual. In the proceeding the starch in the rice is converted by means of certain micro-organisms Chlamydomucor oryzae into sugar and then forms a suitable habitat for the reproduction of yeasts which are probably present in the Raggi but may find their way into the Tapej from other sources. About 100 lbs of rice are used to pitch 1,000 gallons of wash."

ADDENDUM A

After the first edition of this book was published a few common problems of alcohol powered vehicles have come to light. Instead of problems, perhaps it would be better to refer to these as some of the natural characteristics of alcohol. Largely, the area of concern has to do with alcohol's natural cleansing action on an engine. If an engine was run on alcohol from the very beginning there would be no problem. However, after running for an extended length of time on gasoline there will be a buildup of foreign matter in the form of scale in the fuel tank and fuel lines. After switching to alcohol these deposits may break loose and sometimes clog the jets and or other small passage ways in the carburetor.

The second area of concern is the engine itself. A normal gasoline engine contains deposits of carbon. If this same engine was converted to alcohol the natuaral cleansing action of the alcohol would start to loosen up the carbon. Most of this carbon will burn up or pass through with no ill effect. However, there is a possibility that a particle of carbon will stick in a valve or between the spark plug points, causing a rough running engine.

Another characteristic of alcohol is its resistance to vaporization at cold temperatures. For good cold weather running some method of preheating the alcohol is required. It is beyond the scope of this book to get into specific designs. However, since the average automobile engine produces enough heat to warm a seven room house, the mechanically wise should be able to capture and divert some of this heat to warm the alcohol.